RHYTHMS OF LIFE

RHYTHMS OF LIFE

*The Biological Clocks that Control the Daily
Lives of Every Living Thing*

Russell G. Foster & Leon Kreitzman

Yale University Press
New Haven and London

To my wife Elizabeth and children, Charlotte, William and
Victoria, and to my Mother, and the memory of my
Grandparents George and Rose Dixon (RGF)

To my wife Linda and children Sophie and Leah (LK)

First paperback edition published in 2005
First published in the United States in 2004 by Yale University Press
First published in Great Britain in 2004 by Profile Books

Copyright © 2004, 2005 by Russell G. Foster and Leon Kreitzman

Printed in the United States of America.

Library of Congress Control Number: 2004105609

ISBN 0-300-10574-6 (cloth: alk. paper)
ISBN 0-300-10969-5 (pbk.: alk. paper)

A catalogue record for this book is available from the British Library.

The paper in this book meets the guidelines for permanence and durability
of the Committee on Production Guidelines for Book Longevity of the
Council on Library Resources.

10 9 8 7 6 5 4 3 2

CONTENTS

FOREWORD

Time is embedded in our genes. Cells are the true 'miracle' of evolution, for they are the basis of life and among their amazing abilities they can tell the time. Biological clocks can be found everywhere, from simple bacteria through to worms, birds and, of course, us. The reason for this expanse of clocks is clear: all life evolved and lives on a planet that rotates on its axis once a day, and so is exposed to large periods of day and night, light and dark. We humans spend about a third of our lives asleep. For the average person this amounts to more than twenty years spent in a horizontal position. No other human activity takes up such a large part of our lives.

But sleeping is not the only process regulated by our biological clock. Most of what happens in our bodies, our physiology and biochemistry, is rhythmic, showing strong day–night differences. Heart beat and blood pressure, liver function (including the important ability to metabolise alcohol), the generation of new cells, body temperature and the production of many hormones all show daily changes. Yet we only notice such things when we lead a 'modern' life – fly to different countries and experience jet-lag or are forced to do shift work. Both are becoming ever more common, and yet the consequences go largely ignored. It is no coincidence that most major human disasters, nuclear accidents like Chernobyl, shipwrecks or train crashes, occur in the middle of the night.

The medical implications of having a biological clock are profound. If most of our physiology alters between day and night, it is hardly a surprise that the action of drugs will vary depending on the time of day that we swallow or inject them. This is an emerging area of medical treatment, but still somewhat under-exploited. A small percentage change in the success of treating cancer, simply by optimising the time of treatment, could lead to the survival of thousands of people who would otherwise die. And we don't even have to develop new drugs – just use the ones we have already in a better way. The same applies to heart disease, diabetes and so on. Even some forms of depression may be associated with disturbances in the biological clock, which can occur in the short, dim days of winter.

Understanding how a biological clock works is not easy to investigate in people. And as with problems in the rest of biology, one requires model systems, that is, animals and plants, that give us clues to how things work. Research on flies, mice and fish, as well as many other more unusual creatures, have given us insights into how nature has made a clock 'tick' and managed to set it properly to local time.

Rhythms of Life provides the reader with a valuable overview of what we currently know about biological time. From stories of human clocks to more recent scientific discoveries of how clocks work in other animals, we can see how specific genes and their products lead us, and our organs, to tell the time. Advances in genetics and molecular biology have taken us a long way in a few years, and our understanding of the clockwork at the molecular level is progressing rapidly.

Professor Lewis Wolpert

ACKNOWLEDGEMENTS

Books about time seem to take a long time to produce. This one is no exception. The harder we tried to clarify the concepts and shorten the text, the longer it took. That we finished it is due to the forbearance of some and the contributions of many.

There have been a large number of people who in one way or another influenced the contents. Among them are Michael and Shirley Menaker, Brian Follett, Ignacio Provencio, Robert Lucas, Mark Hankins, Till Roenneberg, Martha Merrow, Josephine Arendt, Willem DeGrip, Derk-Jan Dijk, David Whitmore, Carl Johnson, Ron Douglas, Fred Turek, Woody Hastings, Andrew Loudon, and all the many colleagues at Imperial College.

William Hrushesky contributed to Chapter 13 on chronotherapy. We drew on the University of Connecticut's excellent site on Biological Rhythms for the chapters on homeostasis and also the search for the clock (http://predator.pnb.uconn.edu/beta/titles/courseref/html). As a rule we have not cited web sites for the simple reason that there is no guarantee as to their longevity, but we have made an exception in this case.

Ben Lobley read the entire text and made many valuable comments. Gareth Williams and Elizabeth Foster provided editorial support. Stuart Peirson assisted with some of the illustrations.

Despite the best efforts of our publisher, Andrew Franklin, and his team at Profile Books, there are bound to be some errors. They are entirely our responsibility.

Forbearance was shown by our families and friends. Thanks are due from RGF to Elizabeth, Charlotte, William and Victoria for putting up with the trials and tribulations of writing. Linda, Sophie and Leah have been through it before, not that it gets easier, and LK wants to thank them again for their support.

INTRODUCTION

We eat when we are hungry, sleep when we are tired and drink when we are thirsty. Or so we think. But it is only a thin veneer of civilisation and the alarm clock that gives us the pretence of choice. Left to our natural devices, we would eat, sleep and drink along with many more biological functions, not when we decide to but when the biological clock inside us tells us to.

It is not only our physical behaviours that are dictated by this tyrannical timekeeper. Our moods and emotions also swing in time to a daily rhythm. Even our most intimate moments are subject to the chronometer – the favoured time for making love is 10.00 p.m.

Humans have broken many links with the natural world. Our food comes prepacked, our drink prebottled and we take pills instead of chewing leaves. Electricity turns our nights into days, and central heating our winters into spring. But go deep into a dark cave without a watch and after a few days out of the sunlight we revert to ancient patterns. Deprived of time cues, our rhythms slowly drift out of alignment with the outside world.

The natural world is full of daily, monthly and annual rhythms. The early bird catches the worm. Dormice hibernate away the winter. Plants open and close their flowers at the same time each day. Bees search out nectar-rich flowers as though by prearranged appointment. Palolo worms swarm once a year in time with the lunar cycle. And the grunion fish adds in a tidal timer for good measure. Just as we humans are time-dependent creatures, so other organisms have been using an array of timing devices to coordinate their actions in much the same way as we now rely on clocks and calendars to tell us when to do whatever.

1

The big difference between us and other living things is that to some extent we can cognitively override these ancient hard-wired rhythms. Instead of sleeping as our bodies dictate, we drink another cup of coffee, turn up the radio, roll down the car window and kid ourselves that we can beat a few billion years of evolution.

We cannot. We and just about every living thing on the planet – animals, plants, algae, bacteria – have a biological clock that was first set ticking more than three billion years ago.

The climate changes; mountain ranges form and continents are remodelled. But the one constant in this ever-changing environment, since the earth and the moon locked into their orbits, is that the earth will turn on its axis, within a minute or two, every 24 hours; that every 365.25 days, Sirius the Dog Star will rise with the sun; the moon will wax and wane every 29.5 days; and twice a day the tides will roll over the shore. It is small wonder that these basic rhythms are etched into living creatures; and small wonder that the ability to anticipate and exploit these changes has an evolutionary advantage. The rhythms that are a consequence and a necessity for living on a rotating planet are time markers. Through these internal timing processes, organisms have adapted to maximise their chances of reproduction in a temporal environment that changes daily with unfailing regularity.

These rhythms are generated within us. Timing in living organisms controls a wide range of behaviours. Infradian (longer than a day) rhythms time behaviours with periods of months or a year, such as migration, reproduction and hibernation; some rhythms are even longer, such as the reproductive cycle of the cicada, which climaxes over a two-week period every 13 or 17 years.

Daily circadian rhythms (*circa*, about; *diem*, a day) are orchestrated by a central clock to keep our bodily systems working in harmony. When this internal timer is disrupted we suffer from the relatively mild symptoms of jet-lag through to serious and potentially life-threatening conditions such as depression and sleep disorders.

Like the conductor of an orchestra, the clock keeps the ensemble of the human body beating to a collective time. It keeps everything from happening all at once and ensures that the biochemistry of the body runs on time and in order. Biological clocks synchronise the times of activity and rest of both diurnal (daytime) and nocturnal (night-time) organisms and those that are crepuscular (active at dusk and dawn), to ensure that peak activity occurs when food, sunlight or prey is available. They enable us and other living things to anticipate the predictable rhythmic changes in our environment: light, temperature, humidity and ultraviolet radiation.

Biological clocks impose a structure that enables organisms to change their behavioural priorities in relation to the time of day, month or year. They are reset at sunrise and sunset each day to link astronomical time with an organism's internal time in the same way that a signal from a radio station can be used to reset a wristwatch to the invariant oscillating of an atomic clock.

These rhythms are studied for many reasons. They are interesting in their own right as biological mechanisms. They arouse the curiosity of field biologists, who are interested in the way in which animals and plants adapt to the environment in which they live. Physiologists and molecular biologists have been taking the clocks apart and working out the clockwork mechanisms. Biological clocks provide key insights in studies on the genetics of behaviour. The best-characterised models we have of behaviour are controlled by circadian rhythms. Genes, proteins, and neurotransmitters that together account for complex behaviours such as the timing of sleep have been identified and anatomically located.

And they are of great interest to clinicians. There is considerable evidence that the way we treat cancers and many other diseases and disorders could be improved if we understood more about the daily temporal profile of the disease and the importance of getting the timing right in regard to the drugs used to treat it.

Any book about biological rhythms will be incomplete. The subject has exploded in recent years and, as with any book about a scientific subject, this one carries a sell-by date. Apart from the flood of new information, we have the problem of what to leave in and how much to leave out. There is also the issue about whether we concentrate on the science, the scientists or the biological principles.

To provide some narrative coherence, we have adopted a mildly chronological tale. It is not too rigid, but we feel it is important to acknowledge that modern biology is built by the laying down of many small bricks by many hands. Although the theory of evolution is biology's powerful unifying principle, it is an explanatory rather than a predictive theory. Biology is still an experimental and observational science made up of many small and seemingly disparate insights.

There is no need to be a biologist to read this book, but some understanding will help. We have tried to keep in mind the mythical 'intelligent lay reader' who has no biological training but no doubt in places we have assumed too much. For that we apologise in advance, as we do for the unavoidable use in places of terminology that is daunting when unfamiliar. Some pages may be hard going, as there is always a conflict between the demands of accuracy and accessibility. Compression is the enemy of truth and sometimes the description is, unfortunately, as simple as it can get. Short of wrapping a wet towel round the head, the other option is to skip some pages. No irreparable damage will be done.

But we also have to write for the expert, the student and the informed professional of other disciplines. Riding all these horses at once is always something of an insuperable obstacle. There are only so many pages. We have left aside a raft of rhythms – intertidal, lunar, ultradian rhythms of less than a day, and the one-off rhythms of birth and death. We have tried, however, to be scrupulously accurate within the bounds of current understanding even if we are not complete.

It would be a pity if the reader gained the impression that circadian rhythm research has been overwhelmingly on mammals, birds and insects. Huge amounts of vital work have been done with bacteria, plants, algae, crustacea, fish and reptiles. Our bias is in no way intended to marginalise this work but merely reflects yet again the constraints of time and the number of pages.

We hope that both the diligent and the casual reader will finish this book with at least a passing understanding of the basics of biological clocks and some sense of the nature of biological research. Rossini said of Wagner's operas, 'Mr Wagner has lovely moments but awful quarters of an hour.' Biological research is like that: weeks, months and even years of tedium are punctuated by the undiluted thrill of finding something new. Biological understanding is not yet sufficient to describe this emotional state, but what drives many researchers are those 'lovely moments'.

Many of the scientists involved in research on biological clocks are mentioned in this book; even more are not. Today there are probably well over a thousand scientists working on the basic science of biological time. At least 10 times as many are working on applying this information in medicine, agriculture, horticulture, manned space flights and warfare.

In all industries, round-the-clock activity is now the norm. People

4

employed in the utilities, process industries, transport, manufacturing, finance, leisure, retailing, emergency services, media and education work to the beat of an artificial rhythm. All of us in the developed world now live in a '24/7' society. This imposed structure is in conflict with our basic biology. The impact can be seen in our struggle to balance our daily lives with the stresses this places on our physical health and mental well-being. We are now aware of this fundamental tension between the way we want to live and the way we are built to live. It is hoped that our developing understanding of the basic biology will provide us with a means to resolve this fundamental dilemma of modern living.

Biological rhythms have fascinated writers and poets for thousands of years. Chapter 1 sketches in the background to these rhythms, particularly the daily (circadian) cycle, and introduces not only some of the differences between them but also a key protagonist of this book. Much of what we know about the genetic basis of behaviour comes from the study of *Drosophila*, the fruit-fly; one of this tiny animal's keenest students was Colin Pittendrigh, who made biological rhythms his life's work. Pittendrigh talked of the external environment in which sunrise was followed by sunset and then sunrise again as the Day Outside, while an organism's biological clock controlled the Day Inside. A major theme of this book is how the two Days are inextricably linked.

Asked to name an animal that is industrious and sociable, most people would put the bee near the top of the list. In Chapter 2 we explore what has been called the bee's sense of time, most notable is its ability to tell its hivemates the direction and location of a food source using bearings from the sun's journey across the heavens. If bees do it, so do birds and many other animals have an uncanny ability to read an internal clock that locates them just as securely in the temporal dimension as their other sensors provide their spatial awareness.

Chapter 3 deals in definitions. What do we mean by a clock and how do we recognise one when we find it? All clocks must have an oscillator that produces a rhythmic beat, but what is an oscillator and how are oscillations produced? The 'hands' are the regular effect, or the physiological rhythm that denotes the existence of the internal beat. But to turn this into a clock that produces time there has to be a link with the predictable daily cycle of the earth's rotation. For most of us it is the bleep from a radio station that connects the Day Outside to the Day Inside. This is not an option for other living creatures: they have to use more direct methods such as dawn and dusk. Furthermore, to be a useful time-keeper a clock has to be unaffected by temperature. This was a matter of some debate between the early researchers.

The empathic relationship with the predictable regularity of the external world, going with the daily flow rather than fighting to subdue it, marks a key difference between the circadian and homeostatic approaches to equilibrium. Chapter 4 examines how organisms have adapted and 'go with the flow' as they time their biology to their local environment.

From the early observations by Curtis Richter that rat behaviour is rhythmic, it took over 70 years to locate the mammalian master clock finally to a cluster of about 20,000 cells in the anterior part of the hypothalamus in the brain. This small group of cells revels in the long name of the suprachiasmatic nuclei (SCN). Chapter 5 is the story of this search, which is a model of biological investigation and interesting in itself as an example of the painstaking nature of intellectual discovery. In the past two decades it has become increasingly difficult to speak solely of 'the' biological clock and we now talk of circadian systems, as it is becoming clear that although there may be a central clock in some species, in most species time is distributed throughout the organism.

Chapter 6 concentrates on the way in which light is the principal agent that entrains the internal clock mechanism to the external cycle of the sun and the stars. We are used to rods and cones as the light sensors in the eye, and so are most biologists, so it was something of a shock to find in mammals a third mysterious photoreceptor with its own dedicated neural pathway that connects the SCN to the world outside. The knowledge of that new photoreceptor is already being put to practical use in the treatment of people suffering from certain eye diseases.

In many ways, Chapter 7 is the heart of the book. It is also the most difficult, and whereas some may regard that as a challenge, most readers are advised to take it in bite-sized pieces. Essentially, the chapter describes how the rhythmic control of a particular behaviour, such as locomotor activity in a fruit-fly, can be characterised at the molecular level. It begins with ground-breaking research by Seymour Benzer and Ronald Konopka, who discovered that just as single genes can affect physical characteristics, a single gene can influence behaviour. There are still many gaps in the story, but essentially we can name molecules, locate them in space and explain how the concentration of a fruit-fly's activity at specific times of the day is the result of the interaction of a bunch of genes and proteins. This work has led, as we describe in Chapter 11, to the discovery in humans of the first single gene to influence a specific behaviour.

The focus of this book is on mammals, but in Chapter 8 we compare their circadian systems with that of birds, insects, fungi and tiny bacteria. There is a wide diversity of circadian rhythms but the fundamental mechanisms that generate them are similar across the living world. This is not immediately obvious. Birds, for example, have clocks in their eyes and the *Limulus* crab has its photoreceptor in its tail, but despite this, common patterns can be discerned.

Apart from the rhythms of daily living, life on the planet is also marked by six-monthly and annual events such as migration, hibernation and reproduction. The change in daylength, or perhaps night length, is used as a photoperiodic signal to ensure that the all-important breeding takes place at the most propitious time, namely when there is plenty of food about. Chapter 9 describes how photoperiodism works – or at least what we know – and introduces the circannual clock, another means for an organism not only to tell the time but also to keep a diary.

Despite their ubiquity and their likely appearance in living organisms over three billion years ago, in Chapter 10 we outlined how, in their evolutionary history, circadian systems have been developed on at least four separate occasions. Evolutionary studies are, by their nature, mainly speculative, but biology can only be seriously considered in the light of evolution, and thinking of what might have been is a powerful indicator of what might be.

In Chapter 11 the emphasis switches from fruit-flies, rodents and house sparrows to humans. We radically challenge our natural environment, creating one that is deemed to be better suited to our needs. In the process, we often attempt to impose a new way of living on a biological pattern that has evolved over millions of years. There are real clock genes influencing real circadian behaviours in humans. One has even been found. Perhaps the most important human behaviour that is intimately linked with the circadian system is sleep. In this chapter we describe how sleep is not a default state but a managed behaviour, and how mental alertness and physical performance are also part of a circadian control system.

Chapter 12 details some of the things that happen when our clocks go wrong. Seasonal Affective Disorder, or winter blues, affects millions of people and makes winter a miserable time. There is growing evidence that circadian malfunctions are involved in other depressive illnesses. Schizophrenics, and people with bipolar disorder, have difficulties with timing activities and this may be a symptom related to a circadian defect rather than dysfunctional behaviour. Shift workers on circadian-unfriendly rotations may well have shift-lag, which is far worse in its long-term effects than the discomforts of jet-lag.

Penultimately, in Chapter 13 we discuss chronotherapy, or the application of circadian principles to medical practice. There is considerable evidence that the efficacy of cancer treatments could be improved if drug administration were delivered in accordance with the circadian biology of the patient, which not only affects the characteristics of the disease but also has an effect on the delivery and efficacy of the drug. Other diseases and medical conditions are susceptible to chronotherapeutic practice, but while it is undoubtedly beneficial in some cases, it is not yet proven to be of use in every case.

Human beings are studied at three levels: the biological, the psychological and the sociological. Time cuts across all three domains. At base, our biology still determines when we do what best. As individuals we have differing perceptions and attitudes to time.

Le comte du Nouy in the 1930s believed that women have a fundamentally different perception of time from men. He wrote:

Western man makes clocks with smaller and smaller divisions until he can now measure a millionth of a second. He assumes that the measurement of a fraction of a second represents an absolute measure of some strictly objective reality. A woman's sense of time is somewhat different from a man's, and the two divergent senses are cause of not a little confusion and sometimes friction. Her sense of time is not fractional or length oriented, but event *oriented.*

He reasoned that this results from the various cycles that regulate a woman's experience throughout life, most of which are not experienced by the male. These cycles are essentially related to child-bearing, puberty, monthly periods, gestation periods, menopause, and so forth. The result is that a woman is timing life, not by the even spacing of the minutes or the hours in the way that a man times his, but in cycles that are much longer and not nearly so precise. This is the old dichotomy between 'natural' time and 'clock' time.

Du Nouy also wrote (du Nouy, 1937):

The intervening time spaces are not attended to in the same way. When a woman responds to her impatient husband as he waits to take the family to the theatre, by saying 'Coming, dear, right away,' she does not mean this literally. She means only that at that moment this is the next event she has in mind: to join her husband. Meanwhile, he makes a mental note of her reply and allows her forty-five seconds to make the trip from her bedroom to the front door! Consequently, he is frustrated when, ten minutes later, he is still pacing up and down the hall ... Neither party seems able to accept the other's sense of time. And children have the same problem with grownups.

Basic biology, gender differences or a cultural construct? Take your pick. Du Nouy's analysis is unsupported by hard data, but leaving aside the chauvinism, there is little doubt that we all have different approaches to time and how we use it, and that much of this is dependent on our personalities.

But time is a social construct. We are taught that time is valuable and that procrastination is the thief of time. We live in a world where few of us feel that we have the ability or even the inclination to accede to the request 'don't just do something, sit there'. In our modern society, everybody not only has to be busy, but also has to be seen to be busy. All the time.

We have to make choices. We are diurnal creatures but we live in a 24-hour world. We can manage the continued development of the 24-hour society and if necessary use pharmacological intervention or light-based therapies to counteract the biological downside of working at night; or we can reject the continuing trend and attempt to reverse the breakdown of the traditional temporal structure of our lives. The choice, as ever, is not completely free but it is one that we have to make.

1

THE DAY WITHIN AND
THE DAY WITHOUT

A rose is not necessarily and unqualifiedly a rose; that is to say, it is a very different biochemical system at noon and at midnight.

COLIN PITTENDRIGH (PITTENDRIGH, 1993)

The first thing most of us think on waking is 'what time is it?' Clocks rule our lives. They instruct us when to sleep, wake, work, play, eat, drink and pray. In our modern world we need to know the time to tell us what to do.

Yet the clocks with which we are familiar are unnatural: recent human inventions, machines whose products are hours, minutes and seconds. We now take our time from the energy states of an electron in the caesium atom. There is no longer any connection, so we believe, between our clocks and the rhythmic cycles of nature – the dawn and dusk; the lengthening days and shortening nights reversing as autumn and winter draw in; the waxing and waning of the moon and the rising and falling of the tide.

But we would be wrong. Despite electricity and atomic clocks, our bodies still beat to a daily cycle. We do not recognise it for what it is because we live now in a world beset with all manner of artificial timing cues so that our basic internal clocks are often 'masked'. But if we travel in a jet plane across a few time zones, we are soon aware that it is not so easy to beat our biology.

Isolating the individual from the environment soon reveals these internal daily rhythms. Human volunteers have gone deep underground and stayed in a constant light environment for weeks on end. With no way of knowing day from night, their body rhythms started to drift out of synchronisation with the outside world. After about a fortnight they were going to bed around what was in fact midday and rising at around eight in the evening. After about a month they were back, more or less, in synchrony with the outside world, before drifting off again.

Just about everything we humans do shows these circadian (*circa*, about; *diem*, a day) rhythms. They keep everything running like clockwork, and assign a time for everything. This programmed regularity is vital as a means of stopping everything from happening at once. If we did not separate our bodily events by time we would be in a fine mess. For example, it helps if we do not need to urinate while we sleep. Kidney function and hence urine production are reduced at night as our urination and sleep cycles are out of phase.

Other rhythms include our body temperature, which is higher during the day than at night, as is our heart beat and blood pressure; we have a clear sleep/wake cycle; our cognitive abilities change rhythmically over a 24-hour period. Tooth pain is lowest after lunch; proof reading and sprint swimming are best performed in the evening; labour pains more often begin at night and most natural births occur in the early hours; sudden cardiac death is more likely in the morning. Even the strength of a handshake varies (Appendix I).

These variations in our physical, emotional and cognitive performance are not trivial. Depending on the task, the performance change between the daily high point and the daily low point can be equivalent to the effect on performance of drinking the legal limit of alcohol. The best time of day for doing a given task depends on the nature of the task. For example, complex problem solving or logical reasoning is most efficient around noon. Tasks that rely more on physical coordination, such as athletic events, are performed best in the early evening, around the time of the daily peak in body temperature.

When this internal clock is disrupted we suffer from symptoms that range from relatively mild jet-lag through to serious and potentially life-threatening conditions such as depression and sleep disorders. Large-scale studies show that we are at our most vulnerable in the early morning hours. At the low point of the circadian cycle, the body seems least able to resist cardiac or respiratory difficulties. And it is not merely a coincidence that many of the most dramatic accidents of recent years such as Chernobyl, Three Mile Island and *Exxon Valdez* all happened at night.

It was not always like this. Before sundials were invented, people lived by natural time. The sun, moon and stars determined the pattern of life. The earth turned on its axis and split time into day and night. Its tilt gave us seasons. People rose at dawn, tended their animals and went about their business until sunset. Jesus, as an observant Jew, prayed at daybreak when the cock crowed, soon after noon when the sun was at its highest, and in the evening when three stars were visible in the sky. On cloudy nights, he waited until he was unable to distinguish a blue thread from a black one. We explain this today by noting that in bright light he would have a keen sense of colour vision using the cone photoreceptors in the retina, but as the light faded his visual system would have switched to rod-dominated low-light monochrome vision.

In this world people knew the time by what they were doing. There was a harmony between their daily bodily rhythms and the external world. To use a much-degraded word, there was a holistic conception of time. The constant repetition of day and night, the cyclic patterns of the seasons, planting and harvest, birth and death, were seen as earth-bound patterns of heavenly movements. Plato located time itself as being born in the motion of the heavenly bodies, in the perpetual and unchangeable cyclic motion of the sun, moon and planets.

Of course, it was never that idyllic. In the latter part of the second century BC, the Roman playwright Plautus had one of the characters in a comedy complaining about the tyranny of time and in doing so showed an implicit understanding of the internal rhythms that drive behaviour (Landes, 2000):

The gods confound the man who first found out
How to distinguish hours. Confound him too,
Who in this place set up a sundial,
To cut and hack my days so wretchedly
Into small pieces! When I was a boy,
My belly was my sundial – one surer,
Truer, and more exact than any of them.
The dial told me when 'twas proper time

To go to dinner, when I ought to eat:
But nowadays, why even when I have,
I can't fall to unless the sun gives leave.
The town's so full of these confounded dials.

In one form or another, sundials have been used by different societies for more than 5,000 years. Originally, someone in what was then Mesopotamia (now Iraq) had the notion of pushing a stick into the sand and so plotting the rotation of the earth by following the sun's shadow. They had been introduced in Greece about 50 years before Plautus' character was ranting about them. By the first century BC, Andronicus had turned the primitive sundial into the supremely elegant Tower of the Winds in Athens. This was a sophisticated astronomical observatory that included a sundial, weather vane and compass, and a water clock for cloudy days.

In Plautus' and Andronicus' Greece and all over the ancient world the sun was the time signaller. The Konso of southern Ethiopia marked off the periods of the day by simply pointing to the sun's position in the sky (Aveni, 1990).

But as the sociologist Helga Nowotny has pointed out (Nowotny, 1994):

With the advent of clocks, communally usable, visible reference points
were created, which were superior in accuracy and reliability to the means
of orientation previously used, the observation of heavenly bodies or the
behaviour of animals. It was enough to glance up at the clock on the
church steeple to find out what time it was.

The first mechanical clocks appeared in about 1300 AD. Since then we have been steadily losing a battle with time. Instead of controlling our modern clocks, they control us. We complain that we are always racing against the clock, that we are pushed for time and we are caught in a time trap. The work–life balance is the anxiety of the age, and resolving it will not be possible without some understanding of the underlying biology.

Our detailed knowledge of human circadian rhythms is relatively new, but it has been known for thousands of years that there are rhythmic cycles in plant behaviour. Alexander the Great is supposed to have been fascinated by the tamarind tree opening and closing its flowers in synchrony with the day. In Greek mythology, Chloris was the goddess of flowers and she was helped by the Horae (hence horology), the daughter of Zeus, who controlled the seasons and represented the hours in a day. By the late 1600s, the English poet Andrew Marvell was writing about a floral horologue, a clock made of plants that opened their flowers at different times of the day (Marvell, 1681 [1969]):

> How well the skillful Gardener drew
> Of flow'rs and herbes this Dial new;
> Where from above the milder sun
> Dies through a fragrant Zodiak run;
> And as it works the industrious Bee
> Computes its time as well as we.
> How could such sweet and wholesome Hours
> Be reckon'd with but herbes and Flow'rs.

The idea of a floral clock was formalised by Carl von Linné (Carolus Linnaeus). In 1751, he noted that two species of daisy, the hawk's-beard and the hawkbit, opened and closed with a period that was within a half-hour each day. He suggested planting these daisies along with St John's Wort, marigolds, water-lilies and other species in a circle. The rhythmic opening and closing of the plants would be the effective hands of this clock.

Many plants open their leaves in the day and close them at night. They need the energy of sunlight to drive the photosynthetic reaction that converts carbon dioxide and water into sugars, and in the process release oxygen. Each leaf cell must be exposed to the surrounding air for efficient gas–water exchange and to light for photosynthesis. Plants cannot move and so change their immediate surroundings. If it is very hot, their leaves will droop, but the one thing they cannot do is move into the shade. The circadian clock is vital to ensure that the many physiological activities, especially those related to photosynthesis, occur in the right sequence at the optimal time.

Tobacco plants, stocks and evening primroses release their scent as the sun starts to go down at dusk. These plants attract pollinating moths and night-flying insects. The plants tend to be white or pale. Colour vision is difficult under low light, and white best reflects the mainly bluish tinge of evening light. But plants cannot release their scent in a timely manner simply in response to an environmental cue, such as the lowering of the light levels. They need time to produce the oils. To coincide with the appearance of the nocturnal insects, the plant has to anticipate the sunset and produce the scent on a circadian schedule.

As the moths and flying insects come out, so do the predators. The birds give way to the bats and the night-adapted owl. Nocturnal animals have rhythms that are diametrically out of phase with diurnal animals. Whether diurnal, nocturnal or crepuscular (active at twilight or just before dawn), all organisms have to somehow anticipate the coming sunset and sunrise if they are to be in the right state and in the right place at the right time. Different species occupy the same space, but they are divided by time. One runs on daytime, the other on night-time.

It is the same story in the sea; Ron Douglas (personal communication) writes:

> *The eyes of fish, for example, take about twenty minutes to change from night-time mode to daylight vision. An animal whose eyes are prepared for the coming of dawn will be able to avoid a predator and catch its prey when the sun rises more efficiently than one who simply reacts to the light.*

There are species of fish and insects that live in caves and have never seen the light. Yet even they seem to have retained the circadian rhythms of their ancestors that lived in a light environment. The assumption is that these clocks are used to keep some sense of internal order, stopping everything from happening at once. Some organisms live 5,000 and more metres down near thermal vents at the bottom of the oceans. This is where life may possibly have begun. No light penetrates the oceans below about 300 metres so it is possible that, in the history of the species, none has ever been exposed to daylight. In these dark recesses the organisms may or may not have clocks. We don't yet know. But every creature, even those in the darkest depths, lives on a rotating planet travelling around a sun, and with that seems to go circadian rhythms.

A rhythm is any sequence of regularly recurring functions or events. We talk of the rhythm in a poem; in a painting; in a piece of music. It comes from a Latin word meaning to flow. While the repeat pattern of tiling on a marbled floor or the use of alternating patterns of light and shade in a painting makes for strong spatial rhythms, in the living world it is the regular recurrence of an event through time that we regard as a biological rhythm. These rhythms are ubiquitous, as J. T. Fraser, an American who is regarded as perhaps the leading authority on the philosophy of time, points out: 'Animals and plants that share the same ecological niche must co-ordinate their biological rhythms; there must be a chasing time, an eating time, and drinking, mating and building times' (Fraser, 1987).

Apart from daily circadian rhythms there are intertidal (about 12.8 hours), circalunar (about 29.5 days) and annual (about 365 days) rhythms. Some rhythms are even longer, such as the reproductive cycle of the cicada, which climaxes over a two-week period every 13 or 17 years. Those rhythms with a period longer than a day are known as infradian. Ultradian rhythms have a period of less than a day. The most obvious is the heart-beat, which pulses around once a second. But although it is rhythmic the heart rate is not constant. It changes with demand, as you discover when you run up a flight of stairs. Blood pressure also varies with demand, but like the heart-beat there is an underlying circadian rhythm to our blood pressure, which is at its lowest at about 4.00–6.00 a.m. It rises steadily from then on and peaks in the early afternoon before starting to fall. By contrast, the Mongolian gerbil is nocturnal and its blood pressure is highest at about 4.00 a.m., when it is out searching for food.

The first person to study biological rhythms scientifically was the French astronomer Jean Jacques Ortous de Mairan. De Mairan was interested in the rotation of the earth and was curious to find out why the leaves of plants were rigid during the day and drooped at night in time with this rotation (see Figure 1.1). In 1729, he put a *Mimosa* plant in a cupboard to see what happened when it was kept in the dark. He peeked in at various times and although the plant was permanently in the dark its leaves still opened and closed rhythmically – it was as though the plant had its own representation of day and night. The plant's leaves still drooped during its subjective night and stiffened up during its subjective day. Furthermore, all the leaves moved at the same time. De Mairan had unknowingly identified the first circadian rhythm, although the term would not be introduced for another 230 years (de Mairan, 1729).

This simple experiment was the first to show that daily rhythms were not a response to changing light levels but were internal, or, in the more

16

Figure 1.1. Facsimile illustration from Darwin's book *The Power of Movement in Plants*, showing the rhythmic movement of leaves at two times of day. De Mairan studied *Mimosa* and described that leaf movement continued under constant conditions. He reported his observations to l'Académie Royale des Sciences in Paris in 1729.

formal term, endogenous. It was another Frenchman, Henri Louis Duhamel du Monceau, a pioneer agronomist, who began their detailed investigation.

Duhamel reasoned that the mimosa leaf movements might have been a response to changes in temperature. So he put the plants in the constant-temperature environment of a salt mine. In constant darkness and constant temperature the plant still showed a regular pattern of leaf movement. But the importance of de Mairan's and Duhamel's work still did not register at the time. Neither did a study by the Swiss botanist Alphonse de Candolle in 1832 (de Candolle, 1832). He discovered that the period of the leaf movement rhythm in constant conditions was not exactly 24 hours. It varied slightly from plant to plant. This was further evidence that the clock was internal. If the rhythmic effect was in response to an external cue it would be expected that all organisms would show the same period, in all cases exactly 24 hours.

Charles Darwin took an interest in these findings and designed an apparatus for measuring leaf movements. He published his results in

The Power of Movement in Plants, a book he wrote with his son Francis (Darwin & Darwin, 1880).

But it was not until the pioneering studies of Erwin Bünning in Germany that rigorous investigation of biological rhythms began (Bünning, 1973). In 1930 he found that the leaf movements of the common bean *Phaseolus* oscillated in constant darkness with a mean period of 24.4 hours. Bünning established a salient property of circadian clocks: when kept in constant conditions they run with a period close to, but never exactly, 24 hours. This daily pattern of an organism kept in constant conditions of darkness or light is known as its 'free-running' rhythm. This rhythm is innate and hard-wired into the genome. A light/dark cycle or other cues can synchronise the rhythms but they do not cause them. The free-running rhythm is the ticking of the genetically programmed clock.

Colin Pittendrigh, a British-born biologist who, like Bünning, is one of the fathers of circadian research, was one of several scientists who found that when the fruit-fly *Drosophila* was kept in constant conditions, the free-running rhythm was also not quite 24 hours. If the rhythm was a direct response to an environmental cue, such as the daily cycles of light and dark, tides, electromagnetic field or cosmic ray showers, then it would be reasonable to expect that in all species the clock's cycle would be only a few seconds or so shy of 24 hours. But it is not. It varies between 22 and 28 hours among different species and is pulled into synchrony by a specific signal, usually the rising and setting of the sun. This alone suggests an independent, self-sustaining endogenous clock.

These daily rhythms are very ancient. *Limulus*, the horseshoe crab, is an archaic-looking creature that has been pretty much unchanged over the past 350 million years and is about as unlovable as it gets – unless you are Robert Barlow Jr, a neuroscience professor at Syracuse University. *Limulus* contains about 1,000 photoreceptor clusters, called ommatidia, in its eye; each cluster is roughly 100 times the size of the rods and cones in the human eye. These receptors are some of the largest in any known animal and make *Limulus* a favourite of neurophysiologists.

Barlow's group has shown that there is a circadian clock in the crab's brain that transmits nerve signals to its eyes (Barlow, 1990). The eye's sensitivity varies on a daily cycle even when the animal is kept in complete darkness. The internal rhythm has a period of about 24 hours and at night these signals can increase the eye's sensitivity to light by up to one million times, roughly the increase in sensitivity needed to compensate for the decrease in illumination between full moonlight and midday. When the crab is exposed to natural light, signals from photoreceptor cells in its tail help to keep the circadian clock synchronised with the actual cycle of day and night. It makes sense to have the photoreceptors in the tail, as this is the part of the animal that sticks out of the mud.

J. T. Fraser has listed just some of the species in which these circadian rhythms have been found (Fraser, 1987):

> *Behavioural rhythms have been studied for hundreds of invertebrates from algae through to diatoms, dinoflagellates, corals, sea anemones, jellyfish, parasites that live in the blood, aquatic and terrestrial worms, molluscs, centipedes, spiders, scorpions, mites, ticks and, of course, insects. Equally extensive work has established the presence of circadian rhythms in mammals from Norway rats and deermouse to mice, squirrels, beavers, muskrats, shrews, weasels, kangaroos, sea otters, dogs, cats, apes and humans.*

Some groups of bacteria have been added to the list.

If the common cockroach, *Periplaneta*, is put in a cockroach-size running wheel (Figure 1.2), provided with food and water and left alone in an environmentally controlled cabinet, with the light on for 12 hours and then off for 12 hours, most of its activity each day takes place during the first two to three hours of darkness. At first sight this is not very interesting. It could be that the insect is simply responding to light cues. When the lights are switched off after 12 hours, the insect senses the change in light levels and is immediately active for a period and then goes back to being quiescent. As the late John Brady wrote, 'this is about as unexciting a finding as banging the side of the cage every twenty-four hours and frightening it into activity'. But if the cockroach is then kept in constant darkness, there is a bunching of activity into a two- to three-hour period that recurs roughly every 24 hours. This pattern repeats itself day after day after day (Brady, 1979). Even in the constant conditions of darkness, the cockroach is able to divide time into a subjective 'day' and 'night', and anticipate what in the wild would be night-time. This is interesting, and for scientists half a century ago it was quite fascinating.

In dry, scientific language, when the cockroach was placed in an unchanging or aperiodic environment (constant temperature and illumina-

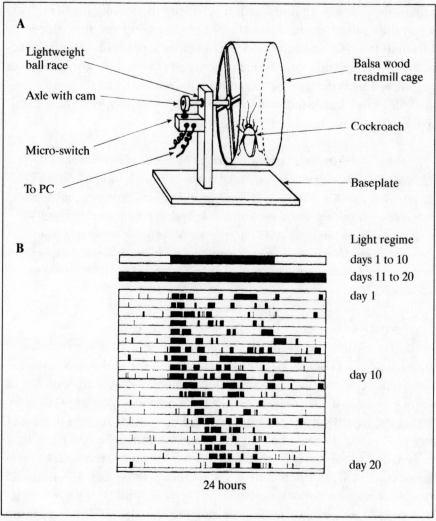

Figure 1.2. (A) A cockroach in a 'running wheel'. Each revolution of the wheel is recorded on the PC and can be represented as a thin vertical black line. (B) This is the activity pattern (actogram) of a cockroach at first exposed to a light/dark (LD) cycle of 12 hours light and 12 hours dark (LD 12:12) (days 1–10) and then left in complete darkness (DD). The black bars and lines denote cockroach activity. On days 1–10, cockroach activity is entrained to the light/dark cycle – the onset of the main activity period is at almost exactly the same time. When left in complete darkness from day 11, the cockroach displays a free-running rhythm (days 11–20) of greater than 24 hours and hence the activity period begins a little later each 'day'. From Brady (1982).

tion) its daily periodicity persisted indefinitely in the absence of external driving factors (Figure 1.2). In lay terms, there is an internal clock that ticks away at close to 24 hours when kept in unvarying conditions, such as constant darkness.

It is not quite a 24-hour cycle, more like 24½ hours, so the cockroach in total darkness would start its subjective day a half-hour later each day. This drift is a bit like the way in which a grandfather clock runs a bit slow or fast and needs adjusting. Released from the obvious light/dark switch that mimics the daily sunrise and sunset, the animal 'free-runs' with a natural period slightly longer than the solar cycle. The free-running rhythm is synchronised each day and forced exactly to the 24-hour solar cycle by light. For most organisms, light is the main time-giver (*Zeitgeber* in German) that keeps the mechanism synchronised to dawn and dusk.

If a biological clock is to be of any use, its period or cycle length has to be more or less constant irrespective of environmental temperatures. This is a problem because biological reactions are subject to the Q_{10} biological rule. Essentially a 10°C rise in temperature doubles metabolic rate, and hence all the biochemical reactions in an organism double in speed. Human physiology works very hard to keep the internal temperature stable at around 37°C, but the cockroach is a cold-blooded animal, or ectotherm, and its internal temperature rises with the external environment. If the clock's period were not temperature-compensated, then a clock cycle of 24.5 hours at 20°C would run at about 12.3 hours at 30°C. A cockroach that was running about in the middle of the day would not last long.

The cockroach clock displays all the key characteristics of all circadian rhythms, in all species. They show a free-running rhythm under constant conditions with a period that is close to, but not exactly, 24 hours; the free-running rhythm can be entrained to exactly 24 hours by an environmental time-giver; and the rhythm, like all good clocks, is temperature compensated.

But if there is a clock driving these rhythms, where is it? What does it look like? How does it work? How does it synchronise with local time? Is there just one clock? Before we can begin to answer these questions we first need to understand something about biological time.

2

TELLING TIME

We should envy the bees who carry an unbreakable watch in their bodies.
KARL VON FRISCH (VON FRISCH, 1953)

Most of us have an instinctive sense of time. If we are out and about we have at least a rough idea of what the time is, even without a watch. It may not be very accurate, but around noon some of us would guess the time at say 11.30, others 12.00 or 12.30 and some perhaps 11.00 or 2.00. But very few if any of us would say that it was 9.00 in the morning or 8.00 at night. Even if we could not see the sun, the rumblings in our tummies would give us some hint. Men could even use the roughness of their beards as an added time cue.

Other animals definitely know the time of day without an external watch. Most bees spend their days feeding, resting, collecting nectar, hive cleaning – everything except having sex – in tune with the daily solar cycle. Their internal clock is reset by the sun, but on a cloudy day a bee uses the pattern of polarised light to deduce the position of the sun. Bees do not need sundials: they carry their clocks with them. In a manner of speaking, bees can not only tell the time, they can tell other bees the time. This time-sensing facility has given the honeybee iconic status in the study of circadian rhythms and animal cognition, and in our under-standing and interpretation of animal behaviour.

Aristotle was the first person to record this time-sensing ability. He noticed that when sugared water was set out some distance from a hive, no bees might arrive for several days, but once one had come others arrived soon after (Aristotle, 2002). Beekeepers have long known that honeybees are intensely social animals that can communicate with each other to organise their complex societies. In 1910, August Forel, a Swiss doctor and naturalist, published a book in which he raised the issue of whether bees had *Zeitgedachtnis* – a memory of time (Forel, 1910). Dr Forel used to have his breakfast outdoors during the warm summer months. The family doctor was a punctual man and he noted, as had Aristotle, that when he finished his breakfast a group of bees arrived to eat the remains. A foraging bee had scented the jam, sugar and perhaps even honey and flown to the hive to tell the others, then they had all come to take the food. What was uncanny was that the following day they were there at his table at the same time, and the day after that, and so it went on. The doctor noted they came at the same time even when he tried a little experiment and did not put out any food. He suggested, 'the bees remember the hours at which they had usually found sweets ... They have a memory for time.'

Ingeborg Beling, a student of Karl von Frisch, the great ethologist who shared a Nobel Prize with Nikko Tinbergen and Konrad Lorenz, published the first study on the time-sense of bees. She showed that bees could be trained to visit a given site at a given time. To test whether this was due to an association with some external environmental cue or was due to an internal mechanism, she and her colleague Otto Wahl carried out a series of careful experiments in which the bees were shielded from external factors. They kept experimental colonies in rooms of constant temperature, illumination and humidity. The effect of the daily rhythm of air ionisation was eliminated with the aid of radioactive substances, and testing in a salt gallery 180 metres below ground eliminated cosmic radiation.

Whatever they tried, the bees were as punctual as ever. They would come for their food at the right time. But the idea of an internal clock was hard to stomach. Von Frisch took the view that 'we are dealing here with beings who, seemingly without needing a clock, possess a memory for time, dependent neither on feelings of hunger nor an appreciation of the sun's position and which, like our own appreciation of time, seems to defy further analysis' (von Frisch, 1953).

But it is neither a memory nor appreciation of time that does the time-keeping in bees; it is an endogenous clock. Although Karl von

Frisch was acutely aware of a bee's capabilities, especially the timing of their behaviour, he believed that the bees had a fixed memory for a 24-hour period and could not remember significantly different time intervals. The irony is that it was largely von Frisch's pioneering studies that stimulated work on the nature of the endogenous clock.

Von Frisch spent over half his very long lifetime (he died at 96) working out the details of how bees manage to communicate to each other the precise direction and distance of a new-found food source. He discovered that the peculiar figure-of-eight 'dance' that the foraging bee does on its return in the darkness of the hive described not only the direction of the food source but also the distance. On the vertical honeycomb, which von Frisch referred to as the dance floor, the bee performs a 'waggle dance', which in outline looks something like a coffee bean – two rounded arcs bisected by a central line (von Frisch, 1953). The bee starts by making a short straight run, waggling from side to side and buzzing as it goes. Then it turns left (or right) and walks in a semicircle back to the starting point. The bee then repeats the short run down the middle, makes a semicircle to the opposite side, and returns once again to the starting point (Figure 2.1).

The number of waggles and the intensity of the bee's buzzing in each run indicate distance (honeybees in different parts of the world have different dialects). The bee translates the three-dimensional location of the food in the outside world and its reference to the sun into a two-dimensional map on the comb. In the hive, gravity provides the position of the sun ('up' locates the sun), and the direction of the food is given by the angle that the straight-line section of the dance makes with this imaginary vertical line (Figure 2.1).

It took von Frisch and his colleagues a long time to work this out because studying bees is not easy. Sensing order in the seeming chaos of the bees in the pitch-dark of the hive took a natural historian of genius. Tracking the individual members of a swarm was difficult in his day (nowadays bees have tiny barcodes stuck onto them, and an individual can be identified as it exits and returns to the hive by a laser scanner). Von Frisch had noted that the figure-of-eight dance described above was not the only movement in the returning forager's repertoire. It would also do a round dance without waggles, simply moving round in a circle. He thought at first that the round dance and the figure-of-eight dance were due to different types of food. It took 20 years before he was able

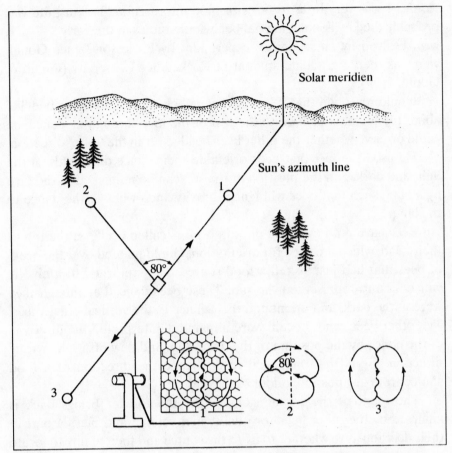

Figure 2.1. Points 1, 2 and 3 are the locations of three different feeding stations. For each, the orientation of a bee's dance is shown. In the hive, gravity provides the position of the sun ('up' locates the sun). The direction of the food is given by the angle that the straight-line section of the dance makes with the sun (vertical line). The number of waggles during the straight line indicates the distance to the food source. From Camhi (1984).

to show that the round dance indicated a food source near the hive, while the waggle dance was for food sources further than about 50 metres.

Even then, his explanation of how the bees communicated through dance was challenged. Other biologists felt that it was more likely that the bees located the nectar-producing plants through odours and the dancing was a by-product as bees 'warmed down' after a flight. In other words, it was very little to do with time information. The dispute was probably the key issue in animal behaviour studies in the early 1970s. It was resolved by an ingenious experiment by Professor James Gould, who was then a graduate student at Rockefeller University (Gould & Gould, 1999).

Gould reasoned that if he could somehow make a foraging bee 'lie' about the location of food to the rest of the hive through its dance, he could observe whether the other bees headed off in the 'false' direction.

He used the fact that bees orientate their dance differently in the light and dark, and the position of the sun can be mimicked in the hive by a lamp such that bees will ignore gravity and orientate the dance to the lamp.

Bees have both eyes and photoreceptors called ocelli on the top of their head, which are used for orientation. Gould painted over the ocelli of bees that had just visited a food source. They returned to a hive in which a lamp represented the sun. These bees danced as though they were in the dark, and orientated their dance to the vertical gravity axis. The other bees, whose ocelli were unpainted, interpreted the direction of the dance by the position of the 'fake' sun, and flew off in the wrong direction. Gould had managed to dissociate the direction of a food source from any possible odour signals.

The scent of the nectar sample that the forager brings back is analysed by the other bees for quality and volume and plays a part in their decision as to whether to fly to the source and their ability to locate the exact spot. But the actual direction they take and distance they fly is communicated through the dance.

The dispute was important for what it said about our understanding of other animals. When von Frisch put forward his ideas that bees not only communicated but also had a distinctive language, he effectively challenged notions about the cognitive abilities of insects (von Frisch, 1953).

After all, bees are simple, aren't they? Bees do not have great intellects in our terms. They are at most a couple of centimetres long and their brains are tiny. A small set of simple rules can explain the sophisticated social behaviour that produces the coordinated activity of a hive. Even though their behaviour looks as though they can solve problems and make contemplative decisions, bees do not think like us. Although there is some evidence that they are able to consider the overall goal of what they are doing when they are searching for new nests, in general they perform mindless tasks. They live by sets of instructions that are familiar to computer programmers as subroutines – do this until the stop code, then into the next subroutine, and so on (Shettleworth, 1998).

Despite this mechanistic simplicity, these humble little bees have an innate ability to work out the location of a food source from its position in relation to the sun; they do this even on cloudy days by reading the pattern of the polarisation of the light, and pass this information to other bees. The bees can communicate how far away the food is up to a distance of about 15 kilometres. And for good measure they can also allow for the fact that the sun moves relative to the hive by about 15° an hour and correct for this when they pass on the information. In other words they have their own built-in global positioning system and a language that enables them to refer to objects and events that are distant in space or time.

It is worth thinking about the enormity of this. It is as though a bee comes back and says, 'Hey, sisters, if you go 6 kilometres in this direction to this spot you'll find lots of food. Flying time is about 15 minutes [bees can fly about 24 kilometres an hour]. It is good stuff and there is plenty of it.' The conditional 'if' is, of course, pure speculation. As Marc Hauser has pointed out, the bee is making a kind of report on a recent event, and whether it goes on to offer an explicit set of instructions is conjecture (Hauser, 2000).

Whereas the dispute in the 1970s was about the means of communication, the chronobiologists are interested in the content of the message. How do the bees know how far they have flown? Distance travelled is time multiplied by speed. Can they tell the time? Do they count wingbeats? Can they estimate distance by sight? Is it a case of communicating 'optical flow', the phenomenon we are all aware of when we travel in a train and look out of the window, and see the nearby landscape flashing past much faster than distant objects? Von Frisch looked at the bee measuring system in windy weather. 'If they encounter a headwind on their flight to the feeding-place, on their return they indicate a greater distance than if there had been no wind; with a following wind a shorter distance is indicated. It seems that their calculation of distance depends on the time required' (von Frisch, 1953).

Honeybees are acutely attuned to the way in which their environment changes with time on a daily basis as they go about collecting nectar and pollen from flowers. They specialise in seeking out one flower species at just the right time. As Professor James Gould explains (Gould & Gould, 1999):

all the information about a flower is time-linked. This makes good sense for bees, since flowers of a given species all produce nectar at about the same time each day, a trick that concentrates foraging on a particular species into a narrow time-window and thus increases the probability of cross-pollination even by very forgetful insects like butterflies.

Bees can be trained to associate a particular scent at a particular time of day with a food reward. After being trained to associate orange blossom, but not lavender, with food at 10.00–11.00 a.m. and then the converse between 11.00 a.m. and 12.00 noon, the bees change their choice at the appropriate time.

The bee pulls off this trick because, like us, it has a circadian clock that is reset daily to run with the solar cycle. The bee can consult this clock and 'check' off the given time and associate this with a particular event, such as orange blossom at 10.00, lavender at 12.00 (Koltermann, 1974). It seems that bees know how to tell the time. It is another matter entirely whether they know that they are telling the time.

In Gould's view a bee's memory 'is organised like an appointment book with each floral rendezvous entered on a separate line'. Von Frisch observed that sometimes a forager would return too late in the day for the rest of the hive to go out for food. Instead of dancing, the bee waited until morning and then did its dance. It not only had to remember the number of waggles to denote distance and the direction of the food in relation to the sun but also had to correct for the fact that it was now some 12 hours later. All this was done inside the dark hive!

It is not easy getting one's head around this. Gould & Gould (1999) have described the challenge to our way of thinking:

When a human decides whether to recommend a restaurant, taking into account its menus, the tastes of the friend being advised, the cost of the food, the distance to the establishment, the ambience of the dining room, the ease of parking and all the other factors that enter into such a decision, we have little hesitation in attributing conscious decision-making to the calculation. When a small frenetic creature enclosed in an exoskeleton and sprouting supernumerary legs and a sting performs an analogous integration of factors, however, our biases spur us to look for another explanation, different in kind.

We find it hard to accept that a bee can tell the time. We are more comfortable with dogs. Sam was a legendary terrier. Each day Sam would run down to the local rail station to greet its owner as the train pulled in at 5.30 p.m. One day the dog arrived as usual but this time his owner did not get off the train. He had had a heart attack at work and had been rushed to hospital. But Sam was there the next day at 5.30 p.m. and the day after that. This went on for several weeks until the owner returned home to convalesce. It was clear that Sam could tell a time, but how? Could a dog have a clock? Or is it a form of telepathy or some weird and wonderful morphic resonance. Of course, the story may not even be true, and may be yet another 'urban' myth. But most people who keep dogs, cats or other mammals appreciate that their pet has at the very least an awareness of time.

It is not just dogs and bees: bats are good timekeepers; so are birds. Oilbirds live in the neotropics in such areas as Venezuela. They are nocturnal and eat fruit that they locate by sight. The birds live in caves, up to 700 metres from the cave opening. Oilbirds navigate in total darkness using echolocation, a similar radar system to that used by bats. Yet despite living in the total dark in their caves, each night at dusk thousands of oilbirds pour forth to search for food, their exit synchronised by a circadian rhythm, entrained to the 24-hour solar cycle by the low levels of light they encounter when they emerge from the cave.

Colin Pittendrigh (Pittendrigh, 1993) described how Gustav Kramer

trained a starling outdoors to go in a particular compass direction for its food reward, evidently using the sun's azimuth as a compass. It was challenged to do so inside a laboratory where an electric light replaced the real sun as direction-giver. In hour after hour, as the bird sought its target direction (where the food should be) it added 15° of arc (counter-clockwise) to the angle it made relative to the artificial sun. Three strong conclusions emerged: (a) the bird knew that the angular velocity of the sun's azimuth is, on average, 15°/hr; (b) it had access to some reliable clock to compensate for its constantly shifting (azimuth) compass; and (c) it knew it was in the northern hemisphere.

In another series of experiments, this time with garden warblers, the birds established a visiting pattern whereby they fed at different feeding stations at different times. In one of the experiments the birds were prevented from visiting any of the stations for three hours. When the visits restarted, the birds went to the correct feed station for the time of day. In other words, they had recognised the three-hour period and readjusted the schedule of visits accordingly (Biebach *et al.*, 1991).

Birds can tell not only the time but also the date. Garden warblers migrate south in the winter. These small birds use the stars to navigate, but that only gives the spatial dimension. The birds also need a temporal cue. They have to know when to fly south, how far they should go, how long they should stay and then when to head back north. If they get it wrong and fly off too early or too late, or fly too far or not far enough, they will miss the other warblers and the chance for good feeding sites and protection in numbers on the southward trip. A warbler who gets the timing wrong on the return journey misses out on the good nesting sites and the best mates.

For their southwards migration in autumn the birds pick up on external cues, such as the shortening of the days, to decide when to migrate. They spend the winters close to the equator. The problem for the birds is that at the equator the days and nights are more or less 12 hours in length throughout the year. The environment is too constant to pick up reliable cues such as changes in the length of daylight.

Eberhard Gwinner of the Research Centre for Ornithology of the Max Planck Society found that garden warblers have two internal clocks. The circannual clock strikes once a year, signalling when to fly back north. Another internal clock keeps it to the 24-hour rhythm that regulates the bird's daily activities (Gwinner, 1996b).

It is hard for us to accept the competences of other organisms even when we confront them. Bees are prewired with at least three separate cognitive modules that enable them to locate nectar by associating odour, colour, flower shape, pattern and time of day with the nectar sources, navigate back to the hive and communicate the navigational information to their sisters. In the bees' world, telling the time is a piece of cake, although it has to be pointed out that some bees are surprisingly bad at using the dance information and finding the right patch. On average, four dance-guided trips are needed to find the site, and the bee may stray far from the intended direction. The advantage is that some of these error-making bees may find new patches.

Although it is clear to us today that the bee has an endogenous clock and has a representation of time, it still raises the question of what we mean when we say that the bee is telling the time. It is all too easy to slip into anthropomorphic thoughts. We cannot ascribe human characteristics to non-human forms, but metaphor is a useful tool, even though it comes with a 'health' warning. The 'pervasive use of metaphors to encapsulate and express scientific ideas (for example, the use of literary metaphors such as "translation," "editing," "reading" for describing molecular processes), while necessary and powerful, also can carry with it the danger of adding false or misleading connotations to the concepts' (Chew & Laubichler, 2003). This is a warning well worth bearing in mind.

Animals and plants cannot tell the time in the way that we do – they do not look at a watch. But they do have clocks inside them. It seems that many living organisms have an internal representation of the solar cycle, and perhaps even a sense of time, but just what do we mean by time, and if there is a biological clock how do we recognise it?

3

OSCILLATORS, CLOCKS AND HOURGLASSES

9,192,631,770 periods of the radiation corresponding to the transition be-
tween the two hyperfine levels of the ground state of the caesium-133 atom
13TH GENERAL CONFERENCE OF WEIGHTS AND MEASURES 1967 –
DEFINITION OF A SECOND

Time has been causing problems for philosophers ever since St Augustine's musings in the fifth century AD. He reflected on the meaning of time and concluded, 'If no one asks me I know, if I wish to explain to one that asketh, I know not'. The problem is that it is difficult, perhaps even impossible, to come up with a definition of time without using the word 'time' itself (Greene, 1999). One answer, apparently favoured by Einstein, is that time is whatever the clock says. Describing time as that which is measured by clocks is somewhat pragmatic rather than principled, but the meaning of time has always been related to its measurement and it is probably going to be the best we can do for the next 1,500 years.

It begs the question, though: what is a clock? And there is not much point in the answer that a clock tells the time. A better shot is the suggestion, admittedly not perfect but nevertheless good enough, that a clock is a device that undergoes regular cycles of motion. The more regular the cycles, the more accurate the clock is.

Brian Greene has pointed out, 'perfectly regular cycles of motion implicitly involve a notion of time, since regular refers to equal time durations elapsing for each cycle' (Greene, 1999), but as this is a book about biology, we can call a halt here and let the physicists and cosmologists earn their crust worrying about it. For us, regular cycles of motion will do.

So what is a regular cycle of motion? Both our breathing rate and heart-beat are fairly regular when we are not stressed. These two pulses probably have been used as rudimentary timing devices since the dawn of mind. Galileo is famously supposed to have used his pulse to time the swing of a lamp while sitting through what was probably a very boring sermon in the cathedral in Pisa. The 19-year-old Galileo noticed that the time it took for an altar lamp to complete its gentle swing (its period) depended on the length of the chain holding the lamp and not on the angle of the swing (the amplitude) (Figure 3.1). The idea of the constant swing of a pendulum was born, accurate clocks became possible as a regular cycle of motion was recognised, and Western civilisation changed.

In Galileo's day, counting his pulse knowing that 70 beats was approximately one minute was about as accurate as using the crude mechanical clocks of the time. Frictional forces made their time-keeping accurate to 15 minutes a day at best. But, unreliable as they were, they were fundamentally different in principle from the sundials and water clocks that preceded them. The sundial tracked the slow motion of sunlight across a dial; the water clock, or clepsydra, allocated time units to the flow of water; King Alfred used the burning of a candle to measure time; and the flow of sand was used in ancient Rome to determine the length of speeches.

The shadow cast by the sun sweeping around a dial directly reflects the relationship between the earth's rotation and its position *vis-à-vis* the sun. It is an analogue device, following the continuous, regular motion of a moving shadow as the earth rotates. In the sundial world, time is continuous and models the heavens. The gnomon, or pointer, is set on a north–south axis so that at midday the shortest shadow falls on this line. The sundial does not generate a regular cycle of motion itself but simply reflects the movement of the sun. There is neither tick nor tock. If the 'hands' are taken away, as when a cloud passes overhead, then time metaphorically disappears. The earth is still rotating, but there is no shadow, no time.

The regular resting pulse used by Galileo, like the beat of a mechanical clock, results not from a continuous analogue movement but instead

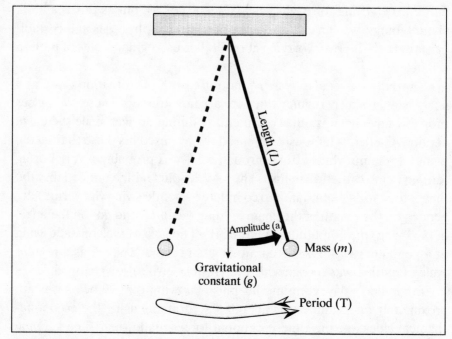

Figure 3.1. A swinging pendulum. *T* (period or *τau*) varies with the length *L* of the pendulum and is independent of the amplitude (a) (also see Figure 3.2).

from the rhythmic oscillation of the heart contracting and relaxing as it pumps blood through the system. The mechanical clocks that had been first invented in the thirteenth century produced a regular rhythm by the translation of the smooth gravitational force of a weight into a digital oscillation. The energy source was a weighted rope wrapped around a cylinder, and a braking arrangement – the escapement – translated the potential energy of the descending weight into an oscillating movement that powered a cog and gear train that drove the hands. David Landes, the pre-eminent historian of the clock, wrote that time was measured by 'something that oscillated at regular intervals rather than something (like sunlight or water) that flowed' (Landes, 2000). In Landes's words:

> *This was the Great Invention; the use of oscillatory motion to track the flow of time. One would have expected something very different – that time, which is itself continuous, even and unidirectional would be best measured by some other continuous, even and unidirectional phenomenon.*

Galileo's genius was his realisation that a constant oscillation of the pendulum had the makings of a far more accurate clock than was then available. Santorio Santorio, a physician friend of his in Venice, began using a short pendulum, which he called a 'pulsilogium', to measure the pulse of his patients. Christiaan Huygens used Galileo's work to make the first pendulum clocks – a classic example of a machine built on the basis of a theory.

Although the pendulum is also subject to frictional forces that gradually decrease the amplitude of the oscillation, the period remains constant, so it can accurately measure the passage of time. If we plot the position of the bob at the end of the pendulum as it swings to and fro, the resulting sine curve gives a measure of the period (*tau*) of the oscillation – or how long it takes to complete one cycle – and the amplitude, which is a measure of the displacement of the bob (Figure 3.2). If the period is measured in degrees, then one complete period is 360°; the phase at a given point is the angle at that point from a given reference, which is usually 0° and might be equivalent in biological rhythm research to, say, 6.00 a.m.

The constant, regular swings of the pendulum beat the time. The escapement mechanism counted these beats by alternately blocking and releasing the wheel train that drove the hands. Tick, tock, tick, tock. Even in today's digital age, the tick-tocking of a grandfather clock is recognised as the audible symbol of the passage of time: the more regular the oscillation, then the more accurate the product. Any clock counts a regular, repeating sequence such as a pendulum swing or an atomic oscillation. Take the back off your wristwatch and look inside. There will be a tiny battery that supplies the energy to keep the thing working. An electronic circuit transduces the energy so that it can modulate and monitor the vibration of a quartz crystal that provides the oscillation. The output is simply a counter.

Because the earth's spin varies fractionally, instead of measuring days, minutes and seconds by the time between one noon and the next, in the 1950s the 'atomic' second was introduced, based on the vibration of a caesium atom. There are 54 electrons spinning around the nucleus of the caesium-133 atom. As the lone electron in the outermost shell spins it creates a tiny magnetic field. The nucleus spins also, creating its own magnetic field. When the two magnetic fields are spinning in the same direction, the total energy within the atom is slightly higher than when the two spins are in opposing directions.

35

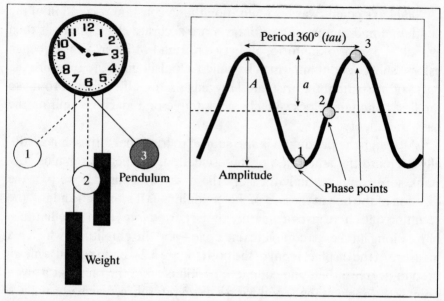

Figure 3.2. In a pendulum clock the descending weight is converted into the oscillating movement of the pendulum. One complete cycle of the pendulum (360°) is defined by its period, or the time it takes to return to its original position after release. The position of the pendulum as it passes through 180° is indicated by three phase positions (1–3). (Oscillation, rhythm, period and cycle are more or less synonymous.) In biological oscillators the amplitude is the peak-to-trough difference (A). By contrast, in a physical oscillator such as a pendulum, the amplitude is the mean-to-peak difference (a) (see Figure 3.1).

When the lonely electron flips its magnetic direction relative to the nucleus it emits or absorbs a tiny quantum of energy in the form of radiation with a frequency of 9,192,631,770 cycles per second. Or rather, we say a second has elapsed when there have been 9,192,631,770 cycles. The atomic second is one 86,400th of the solar day. But although we need precise atomic clocks to synchronise geodesic satellites, sunrise and sunset are good enough indicators for living creatures.

Clocks count regular oscillations. We turn the clock's product into time by matching the oscillations to the solar rhythm. When the sun is at its highest point we say it is noon, and we divide the period of the earth's rotation until the sun is next at its highest into 24 equal parts. Once noon is decided, then the hands sweeping round a dial or numbers clicking through on a digital clock have a sense of direction. Time can be seen to

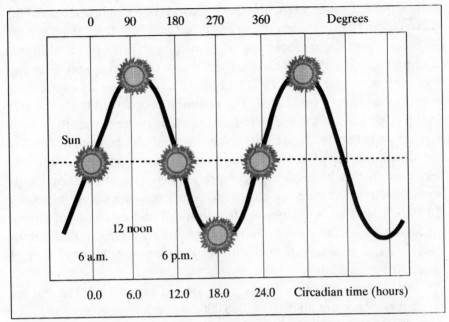

Figure 3.3. The relative movement of the sun over 24 hours or 360°. The dotted line represents the horizon. Organisms have an internal time 'map' representing this regular daily cycle. Human time is a social construction. The week, hour, minute and second do not map to any natural occurrence. But the day and the year are natural rhythms, as is the lunar cycle. The bee maps the daily solar cycle, with its own circadian time.

be moving forward. If the relative position of the sun in the sky to the earth is plotted, it can be thought of as being composed as a series of time points. Each point corresponds to where the sun is at that instant (its *phase* in the parlance) and is an indicator of the time at that instant (Figure 3.3). We measure the phase in hours and minutes, but it is commonly cited in degrees in the scientific literature. Twelve hours is equivalent to a phase shift of 180°. The bee's internal digital oscillator effectively maps this relative movement of the sun and so at any instant the bee 'knows' the time. Of course, it does not 'know' the time in the way we do. It has no concept of time as such. But hard-wired in is a means of tracking the solar cycle.

A regular oscillation is the fundamental requirement for any clock. In the case of the circadian clock the period is about 24 hours. The clock at the heart of the personal computer on which this book is written makes about 2.4 million oscillations a second (hertz, abbreviated as Hz). Slave clocks in other parts of the PC take their beat from their master. Between them these clocks synchronise the activities of the different components. The PC's clock is sustained by a constant supply of energy from the mains or a battery. The biological clock has its own energy sources that keep it oscillating. The common property of the oscillation in the organism's clock and of that in the PC is the tendency to move in a regular manner from one state to another before returning.

Oscillations are ten a penny in nature. The proponents of the eso-teric and frankly impossible-to-comprehend cosmological string theory would have it that even the atoms are composed of smaller units some 10^{-43} cm long. This is way beyond the norms of understanding of anyone without a PhD in cosmology, but if the physicists are right, then forces, particles and all the stuff in the universe depend on the way in which these strings oscillate.

Every atom in our bodies is oscillating at around 10^{16} Hz. Put an-other way, there are 10,000,000,000,000,000 cycles in each second. The period, or length of each cycle, is a reciprocal of this frequency, 0.000,000,000,000,0001 seconds. Pendulums and springs are common-place, and the complex dynamics of linked oscillators, like the familiar Newton's Cradle, have been well characterised. The standard laboratory oscilloscope is a graphic witness to the sinusoidal nature of harmonic motion. A simple electronic system can oscillate continuously and spon-taneously if the inevitable losses of energy are replenished by an anode-voltage feedback.

When it comes to the living world, biology is full of rhythms. The rods and cones in the retina respond to light oscillating at between 10^{15} and 10^{14} Hz. The brain's electrical activity, familiarly seen in the electro-encephalogram, has a frequency of 10^1 Hz, or a period of 0.1 seconds; the heart beats at approximately 10^0 Hz, or once a second; and respira-tion occurs at about one breath every six seconds.

The group of rhythms in which we are most interested are those that beat in harmony with environmental cycles. Circadian rhythms, such as sleep/wake cycles, have a period of 10^5 seconds. Monthly lunar cycles have a period of around 10^6 seconds, and clocks found in the annual moulting of a stag's antlers have a period of 10^7 seconds. At the long end of the spectrum are bamboo plants that flower every seven years, and ci-cada nymphs that emerge as flying insects every 13 or 17 years. Although few century plants (*Agave americana*) take 100 years to flower, 40 or more years is not uncommon, making the longest biological periods of the order of 10^8 seconds. J. T. Fraser has pointed out that the ratio of the fastest to the slowest oscillations in an organism can be $10^{24}:1$ and, as an octave is a frequency range of 2:1, then 'the instruments of the living orchestra extend across 78 octaves' (Fraser, 1987). To play this range would require a piano with a keyboard six metres in length and a pianist with arms three metres long!

Even simple chemical solutions can oscillate of their own accord. In the 1950s, Russian chemists discovered an oscillating chemical reaction when potassium bromate oxidised malonic acid in the presence of suitable catalysts such as manganese and cerium. If the reagents are mixed together in a thin layer in a Petri dish with a pH marker reagent then as the reaction proceeds, the pH marker enables us to see clear waves of colour that radiate out from the centre of the dish. The oscillation period depends on the actual conditions, but at 25°C it is usually around 18 seconds and the show can last well over an hour.

The reaction is nothing less than a chemical clock, although not a particularly convenient one to carry around. It is an example of autocatalysis, in which the presence of a product in the reaction accelerates its own synthesis. The Belousov–Zhabatinsky reaction illustrates two very important points: feedback as a means of providing oscillations, and the ease with which self-emergent complex patterns and structure can be produced from simple components in non-living systems (Prigogine & Stengers, 1984).

Oscillating reactions depend on a positive drive sending the system away from one state and a negative drive forcing it back. The conflicting antagonism of these drives sets up an oscillation. In chemical and biological clocks, this continual antagonism between positive and negative drives results in a continual oscillation. It is a biochemical version of a Push-me Pull-you.

Claude Bernard, the French physiologist, first described antagonistic negative and positive drives as a feature of living systems. He wrote about the continual attempts of the organism to maintain constancy in the internal environment: 'La fixité du milieu intérieur est la condition de la vie libre' (Bernard, 1859). Nicholas Mrosovsky (Mrosovsky, 1990) has pointed out:

here libre means independent. By maintaining constancy in the internal

39

environment we free ourselves of constraints from the external environment; we can live in the desert, in the Arctic, even in outer space. Space travel is the ultimate example of homeostasis, because inside the space vehicle, inside the helmet and the space suit, inside the skin, you find the same old milieu intérieur, 37°C, 90 mg sugar/dl of blood, the same old plasma calcium and potassium concentrations.

Bernard's ideas, particularly in regard to negative feedback, are the basis for homeostasis, the notion that living systems are continually struggling in a constantly changing environment to maintain critical processes at fixed levels, such as the level of sugar in the blood. The resulting appearance of constancy and harmony of the body results from the continual application of negative feedback drives.

But unlike a homeostatic system (see Chapter 4), which is always attempting to return and stay at a fixed point, a self-sustained oscillator like a circadian clock tends in a regular manner to move away from one state to another. Give or take some brilliant engineering, whether it be the natural, molecular engineering of organisms or the mechanical and electrical engineering of us humans, a regular self-sustaining oscillation is the heart of a clock. The only extras required for a simple biological clock are a setting mechanism that links the internal oscillation to the astronomical oscillations of the sun, moon or stars, and a means of making the output biologically useful. Understanding this is the essence of biological circadian clocks. They are oscillators with an output that we measure as biological rhythms, and a resetting mechanism that links them to regular, predictable changes in the environment (Figure 3.4).

This simple model (Figure 3.4), first developed by Arnold Eskin, has been the key metaphor in biological clock research for the past 40 or more years. The overriding question has been: how does a conceptual model such as this translate into biological structures and processes? How can a set of biochemical reactions produce a regular self-sustaining rhythm with a period of about 24 hours that is synchronised with the solar cycle and is temperature compensated? An oscillation is all very well, but how can a biological system, in which temperature-dependent molecular reactions usually occur in fractions of a second, generate a roughly 24-hour oscillation?

Back in 1960, the small band of scientists working in the then nascent field of biological clocks gathered at a defining symposium at Cold

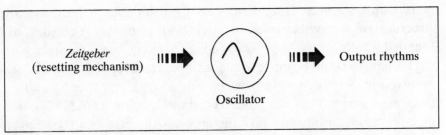

Figure 3.4. A representation of a universal clock, including the circadian sytem.

Spring Harbor outside New York. All the founders were there – Colin Pittendrigh, Erwin Bünning, Franz Halberg, Jürgen Aschoff, J. 'Woody' Hastings, Rutger Wever. Nearly all those present were men, but there were some women, including Pat DeCoursey, who is now the doyenne of the chronobiology community. They did not yet call themselves chrono-biologists, but they did have Halberg's newly minted word for the pheno-menon they were studying. Circadian – about a day – was the new word to describe the near-24-hour rhythms in living organisms.

The 150 scientists present heard Colin Pittendrigh explain the need to bring unifying concepts into the small but fast-developing field. 'We are beset rather than blessed with an enormous number of observations about a great diversity of organisms that range from unicellulars to African violets to man', he said (Pittendrigh, 1960).

The fact that a majority of these observations is highly fascinating is itself a danger – the common danger threatening the biologist – of mistaking ac-quisition of more fascinating facts and more concrete detail, for analytic progress … I take it there is an infinity of facts, relatively few of which are necessary for an understanding of general principles. The student of rhythms protests he has no common mechanism to give his field the unity he would like; the student of insect photoperiodism asserts his system (in-volving eyes and endocrine glands) bears only a superficial functional re-semblance to that of flower initiation by photoperiod. And yet both may be wrong in the sense that there are common mechanisms – built of different concrete parts – in circadian systems and photoperiodic effects everywhere.

When Pittendrigh made his statement there were no genome projects in sight. Rhythm research, like much of biology, was a

41

behavioural science. True, Crick and Watson had determined the structure of deoxyribonucleic acid (DNA), and Max Delbrück had founded what we now know as molecular biology, but these were early days. Changes could be made to timing inputs and reproducible changes occurred in outputs, but in between was a black box that housed the clock mechanism. They had no idea what was inside the box. They could, metaphorically, prod what they thought was the box with their fingers and see what happened. But in truth they did not know where it was in the organism – or, indeed, even if it was a discrete clock.

To grasp the problem facing Pittendrigh and his colleagues, imagine that the various parts of a mechanical clock are laid out on a table in front of you. There are cogs and weights and chains and various other items. If it was a simple clock and you had a curious mind and a few hours to play with, there is a reasonable chance that you could put it all together and get the pendulum swinging. You would have assembled an oscillator that produced a steady, regular beat. To get it to tell the time you would adjust the pendulum length so that it ticked to Universal Time, and set the hands accordingly. It would be a fairly easy exercise because you have a cognitive map of a mechanical clock inside your head. You know roughly how it works and what goes where.

The biologists of the 1950s and 1960s could measure obvious rhythms and so they had an idea of the outputs. They could play around with the inputs and reset the hands of the biological clock by changing the lengths of the light/dark cycles. They could stop the clock by administering various chemicals. They studied the effects of temperature, but they had no idea about the molecular equivalents of the cogs, weights and chains nor the faintest clue about how the inputs and outputs linked to the central mechanism. They could not take the back off the watch. There was no cognitive map to build a biological clock.

Although nearly all of those present at Cold Spring Harbor accepted that an endogenous oscillator was the heart of the circadian clock, there was one important dissenter from this basic notion. Frank Brown Jr, a professor at Northwestern University in Chicago, believed that biological rhythms could be explained by an external, geophysical source. He considered that known geophysical forces, such as magnetism, electromagnetism and even cosmic radiation, were the driving agents of these rhythms. He even postulated as yet unknown forces.

Apart from circadian rhythms, there are many organisms that are entrained by tidal rhythms and others tune in to a lunar-driven cycle, but to qualify as a bone fide clock a biological oscillator has to pass three tests. Does the rhythm persist in the absence of any external time cues, in other words is it self-sustaining? Is there a mechanism for resetting the clock to environmentally critical external cues such as light, dark or temperature? Is the periodicity of the clock temperature compensated; that is, does it hold its rhythm steady at different temperatures?

Brown knew the powerful arguments for an endogenous oscillator. But he found them hard to accept fully because he had serious difficulty with the notion of biological temperature compensation. Pittendrigh had pointed out that for a clock to work accurately, the oscillation could not vary with the inevitable daily variation in temperature. This was exactly the problem faced by John Harrison when he was developing the timepieces with which he eventually won the Royal Navy prize for measuring longitude. Harrison's clocks had to be so accurate that local time could be compared with a reference standard established when a ship left port. By calculating the difference between the local noon time (as determined using a sextant) and the clock's noon time, a measure of the longitude could be obtained because approximately every hour's difference was 15° of longitude. The temperature between embarkation in London in winter and arrival in the Caribbean some six weeks later meant that the ship-board clock had to be temperature compensated to be accurate. Harrison solved the problem by using metals with different expansion coefficients to balance out temperature effects.

This question of temperature compensation was a major issue in the early days of biological clock research (Pittendrigh, 1993). On the basis of an early experiment, Erwin Bünning and Hans Kalmus suggested that in some species of *Drosophila* the circadian rhythm varied with temperature. Bünning had taken *Drosophila* pupae that had been subjected to a 12 hours light/12 hours dark cycle at 26°C and put them in total darkness at 16°C. He used eclosion as the marker for the flies' activity. Eclosion is the term used for the process when the fly emerges from the pupa. There is a strong diurnal rhythm, with nearly all the eclosions occurring around dawn. This has obvious survival value in the wild because conditions at dawn are relatively cool and moist, and the newly emerged adult fly has time to dry its wings and prepare for the higher temperatures of the day.

Eclosion only happens once in a fruit-fly's life, so at first sight it seems strange to use this as a marker of rhythmic behaviour. Similarly, in humans the timing of one-off events such as birth and death is discontinuous. Yet population studies show that human births and deaths follow a pattern, both events peaking in the early morning hours.

Although each individual can add only a single data point, aggregating the data points shows a clear circadian rhythm in the population. In *Drosophila*, those pupae that are not quite ready to emerge at dawn, but would be ready a few hours later, must wait until the next dawn. It is as though a gate that permits emergence (eclosion) opens for a few hours only and then shuts again until the next appropriate circadian phase.

At 26°C the free-running eclosion rhythm of *Drosophila* has a period of about 27 hours. When Bünning measured the emergence times from the pupae at 16°C he found that instead of appearing about 27 hours after being placed in the dark, eclosion took place some 12 hours later, namely at 39 hours. If Bünning was right, it made nonsense of Pittendrigh's claim that clocks had to be temperature compensated.

Colin Pittendrigh repeated this crucial experiment while on holiday in 1952 at the Rocky Mountain Biological Laboratory near Crested Butte in Colorado. As Pittendrigh described it (Pittendrigh, 1993):

> *the facilities were non-existent until I found a well-preserved outhouse (one-holer) near an abandoned mine shaft at approximately 10,000 feet. It was still erect and by now totally odorless. The walls and door were sufficiently intact that some tar paper and nails procured from Crested Butte made it a useful darkroom. Plyboard transformed the seat into an acceptable workbench. The presence of a small crystal-clear creek a few feet away provided a very stable source of low temperature – and a fine opportunity to fly-fish for trout. None of this would have been useful, however, without the pressure cooker that my wife had brought to ease the task of cooking at high altitudes. When emergence activity within them had begun, some vials of* Drosophila pseudoobscura *were placed in the outhouse-darkroom and others in the pressure cooker-darkroom, which was then anchored in the creek to assure a constant low temperature. Its ventilation was effected by attaching, as a snorkel, a black rubber tube to the lid's steam outlet. To minimize distraction from trout, I limited observation to the emergence peak in both darkrooms after two days in darkness. To my surprise, because I was pessimistically expecting to confirm Bünning, the peak in the very much colder pressure cooker was only a couple*

of hours later than that in the outhouse: no more than an hour or so's delay in each of the two cycles at low temperature!

Crude as the experiment was, its outcome was clearly in conflict with Bünning's report of significant temperature dependence. I well remember the excitement of that afternoon in 1952; one had a bear by the tail; the angular velocity of at least one endogenous daily rhythm was, indeed, sufficiently unaffected by temperature to render it a useful clock.

Pittendrigh repeated the experiment again later in his laboratory at Princeton. This time when the culture vials were transferred from a light cycle (at 26°C) to constant darkness (at 16°C) emergence activity was assayed hourly. The initial finding showed that Bünning was indeed correct: the first emergence peak at 16°C, due approximately 27 hours after entry into darkness, was 12 hours overdue.

Pittendrigh went home disconsolate and unable to explain the Colorado data. Fortunately his undergraduate assistants did not give up and continued the experiment through the night. They found the second peak in darkness at 16°C was only two hours later than that at 26°C, the same as in Colorado! Moreover, as the following days showed, the interval between all subsequent peaks at 16°C was only trivially longer than that at 26°C.

A one-off change that can mask the underlying rhythm had misled Bünning. The temperature shock of transferring animals from 26°C to 16°C disturbed the normal eclosion rhythm for several hours, but gradually the system settled down and the normal rhythm was re-established. The circadian rhythm can be temporarily hidden or masked but, as Pittendrigh's students showed, the underlying beat persists and is sufficiently temperature compensated to serve as a useful clock.

But Frank Brown Jr was not convinced that temperature compensation had been proved. He still preferred an as yet unknown exogenous rhythm for explanation. Fred Turek, a former editor of the *Journal of Biological Rhythms* and professor at Northwestern, knew Brown in the mid-1970s. He recalls (personal communication) that Brown

believed that the Q_{10} rule, which states that the metabolic rate approximately doubles for every 10°C rise in temperature, was a universal law of nature. Brown could not imagine that there were any biochemical systems that would not be affected by changes in temperature. Whereas Pittendrigh *concluded that clocks must be temperature compensated or they are worthless, Brown saw the same data and concluded clocks must not be in living systems since biochemical reactions have to obey the laws of temperature. It seems that the argument between Brown and Pittendrigh became personal and then Brown did not want to back down.*

Turek uses the example of how Brown stuck to his hypothesis despite all the mounting evidence he was wrong, to tell his students that you have to be able to stand by your data not your hypotheses. 'When new data arrive, be ready to change your hypothesis.'

The temperature compensation problem may not have been the full story, however. J. 'Woody' Hastings, who knew Brown and Pittendrigh well, says that 'Brown could not measure the rhythms sufficiently accurately and so he always got exactly 24-hour periods. This convinced him it was an exogenous oscillator' (Hastings, 2001).

In fairness to Brown, who was an honest and decent scientist, temperature compensation in a biological clock was hard to swallow in the late 1950s and early 1960s. Brown wrote (Brown, 1960):

*Since geophysical periods are now known always to be available to the organism, and there exists a simple mechanism to explain the circadian variability in terms of them, then following that precept of logic, the law of parsimony, one need no longer postulate the additional existence of an independent internal timer for the inherited living clocks. **Indeed, such a postulation in this instance possesses the added deterrent that no plausible biological scheme for such a timing system has as yet even been conceived** [our emphasis].*

That was the heart of the problem. There was no plausible biological scheme! Remember that in 1960, Watson and Crick's structure of DNA was a mere seven years old and molecular biology was in its infancy. The nature of the code itself, the mechanism by which amino acids were joined together outside the nucleus to make proteins, and the complexities of promoter and inhibitor genes, were still unknown. It was hard enough to imagine how a series of molecular processes could come up with a regular cycle of about 24 hours. Brown simply could not conceive of any molecular process that would enable a temperature-compensated biological oscillator to work, ergo there could not be one. Pittendrigh did not know either. He also knew that most enzyme-catalysed processes are strongly dependent upon temperature, but he was certain there had to be temperature compensation and that eventually researchers would find out how. The problem has still to be solved some 40 years later. But that is not the point. Pittendrigh's boldness established circadian rhythms as a subject not only worth studying with a possibility of success, but one of enough importance that scientists could risk their careers in its pursuit.

One of the keys to success for a scientist is in picking the right problem to work on. Craig Loehle has identified what he calls the Medawar zone as the place for a scientist to be. Loehle (1990) says:

Solving an easy problem has a low payoff, because it was well within reach and does not represent a real advance. Solving a very difficult problem may have a high payoff, but frequently will not pay at all. Many problems are difficult because the associated tools and technology are not advanced enough. The region of optimal benefit lies at an intermediate level of complexity, what I call the Medawar zone in reference to Sir Peter Medawar's characterization of science as the 'art of the soluble'. These intermediate problems have the highest benefit per unit of effort because they are neither too simple to be useful nor too difficult to be solvable.

It is a useful axiom. Modern science is driven by grant funding, which is directed towards the solvable. Had a young Einstein turned up today and put in a research submission along the lines that he wished to study the nature of space and time by conducting thought experiments in an armchair, supported by some esoteric mathematics, and that his research would last a lifetime, he would not have been funded. Similarly, Darwin would not have got very far had he put in a proposal that, although trained as a geologist, he wished to study the problem of speciation by travelling for five years on a research vessel, collecting every specimen and fact that he could find, and then spend another 15 years thinking it all over before writing a book (Loehle, 1990).

Biological clock research has come along in leaps and bounds in the past half century because funding became available along with the research tools. It was the pioneering efforts of Pittendrigh and his colleagues that made clear there was an important biological problem that was worth studying. These clocks have been developed over billions of year to keep time with environmental periodicities so that, as Rensing *et al.* (2001) put it, an organism

> *can be exactly on time with respect to its physiological adaptation. In addition, the circadian clock probably serves to co-ordinate a sequential order of processes within a cell and an organism ... The circadian clock also co-ordinates interactions between individuals of the same or different species, for example the courtship behaviour of* Drosophila.

Or, putting it more prosaically, biological clocks keep organisms in time and on time.

Temperature compensation is still an unsolved problem but a fair part of the detail of the molecular mechanism that makes a biological clock tick and tock has been determined. Huge strides have been made in teasing out the input pathways that entrain the clock. Progress in biological clocks has been neither easy nor simple but the problem has at least seemed soluble.

4

THE CHALLENGE OF DAILY CHANGE

The times they are a-changin'

BOB DYLAN

While Galileo was sitting in the cathedral in Pisa, using his pulse as a surrogate stopwatch, he was recording an interval of time. But counting his pulse gave him no idea as to *the time*.

Duration counting is common in nature and you do not have to be a genius to do it. Pavlov's dogs could manage it. In one set of experiments, Pavlov delayed offering the food for a fixed period of three minutes but he noted that the dogs started to salivate in the last minute before the food was due. In other words, they had learned how to measure the time delay.

Most of us measure short time intervals in the absence of a clock by simply counting one thousand and one, one thousand and two, and so on. But there is a key difference between this duration timing and the circadian clock system. Hudson Hoagland, an American doctor, had shown in the 1930s that the ability to 'count' time was temperature dependent. His wife was ill and had a very high fever, somewhere around 39°C. Hoagland needed to go to the drugstore for some medication for her. When he got home his wife was quite upset with him for taking so long, claiming that he had been away for hours. In truth, his trip had only taken about 20 minutes. His wife's misperception interested Hoagland. Without telling her what he was thinking, he asked her to

49

estimate the duration of a minute by counting to 60 at the rate of one number per second. It turned out that his wife's estimate of a minute was only 30 seconds long. She did much better as her temperature fell (Woodrow, 1951). Her ability to time intervals obviously varied with temperature, so her interval timer worked in an entirely different way to a temperature-compensated circadian clock.

Animals also have no problem in timing short intervals. If a rat is given the task of pressing a right-hand lever every two seconds for food, and a left-hand lever every eight seconds, it quickly learns the routine once it has been trained by reinforcement of the behaviour. Sara Shettleworth has summarised the important distinctions between circadian timing and its clockwork capabilities and the stopwatch effect of interval timing (Shettleworth, 1998):

> A primary function of the circadian system is to adjust the animal's behaviour to local day and night. Among other things it allows animals to learn when and where food is regularly available. The function of interval timing, in contrast, is to adjust behaviour to important events with duration much shorter than a day. Unlike day and night, the duration and time of occurrence of these events are not predictable in advance of individual experience and they can take on any value.

An hourglass is a quintessential duration timer. When the sand has run from one half of the glass to the other, a given period has elapsed. It may be a minute or an hour, depending on the size of the glass and the speed of the flow of sand. But it can only measure a preset duration.

Interval timers such as a stopwatch or an hourglass have to be reset after each period. Unlike circadian clocks they are unable to establish a self-sustaining rhythm. Randy Gallistal, a behavioural neuroscientist from the University of California, says (Morell, 1996):

> what all this interval timing research shows is that animals are storing the value of variables; that they know how long an interval has lasted and they are writing that to memory, in much the same way that a computer does.

Even when the circadian system is disabled, animals are still able to demonstrate interval timing, suggesting that the systems are independent. However, circadian clocks and unidirectional timers combine to provide sophisticated temporal devices. Watch a bird searching for worms on a patch of ground. The circadian rhythm tied to a predictable sunrise ensures that the early bird is ready to start feeding soon after dawn. But as the bird searches over the ground it has to make short-term decisions. It behaves as if it knows the optimal time to spend in any particular patch before moving on. It is as though it has a built-in stopwatch that enables it to follow a rule of thumb, such as to leave after x seconds on this patch or to leave after y seconds of the last success.

Animals measure not only small periods of time. They can do calendars as well. Squirrels, for instance, know about sell-by dates. Their ability to locate acorns that they have previously buried is well known, though in truth they are somewhat spendthrift, not finding a fair percentage of the ones they buried. They bury not only acorns but also other more perishable food items. The squirrels learn that they need to locate the perishable items in a relatively short period or else they will spoil. So they have a memory of not only the location of each item but also the time at which it should be eaten. In anthropomorphic terms they keep a food-store diary.

The tiny cicada counts years. They have an odd lifestyle that was known long ago to the ancient Chinese. Immature periodical cicadas (nymphs) develop underground and suck juices from plant roots. After 13 or 17 years below ground, depending on the species, mature nymphs emerge from the soil at night and climb onto nearby vegetation or any vertical surface. They then moult into winged adults. Their shed outer skins or exoskeletons are found attached to tree trunks and twigs. The emergence is often tightly synchronised, with most nymphs appearing within a few nights of each other.

Adult cicadas live for only two to four weeks. Male courtship songs attract females for mating, but the females are silent. After mating, females lay their eggs in small tree branches. The female's ovipositor slices into the wood and deposits the eggs. One to several dozen eggs can be laid in one branch, with up to 400 eggs being laid by each female in 40–50 sites. Cicada eggs remain in the branch for 6–10 weeks before hatching. The newly hatched ant-like nymphs fall to the ground, where they burrow 6–18 inches underground to feed.

There is no getting away from the conclusion that cicadas have an internal calendar that can count either 13 or 17 years, depending on the species. Rick Karban of the University of California at Davis conducted a beautiful experiment to determine how the cicadas know when to emerge. He collected some nymphs of the *Magicicada* species that had been underground for 15 years and took them to a climate-controlled laboratory where he let them attach to the roots of peach trees that had been manipulated to blossom twice a year.

The trick worked and the cicadas emerged a year early, suggesting that they had double-counted. Karban believes that the cicadas monitor physiological changes in the trees at the same time as they tap into the roots for food. They sense the changes in the tree as each spring a surge of sugars and proteins flows through the roots.

The experiment does not explain how, in Karban's words, 'the cicadas mark off the years' (Karban *et al.*, 2000). The insects seem to have a 17-year calendar that is accurate to a few days, but nobody has a clue as to how they count off the years and keep track of where they are in the count.

Animals have a wider sensory environment than we do. Many species, from sharks to duck-billed platypuses, are able to monitor their surroundings and detect prey using slight changes in an electrical field. Bats find their way around by echolocation. Homing pigeons seem to use tiny variations in the earth's magnetic field to find their way home. And bees apparently use daily fluctuations in the strength of the earth's field to calibrate their internal clocks (Gould & Gould, 1999). We should not be too surprised that they have evolved sophisticated timing systems.

Although we can measure time intervals, we are not as adept at it as other animals. We use tools instead. We can measure time duration either by an interval timer, such as a stopwatch, or through computation via a clock. In the former, the stopwatch reads out directly the amount of time that has elapsed. In the latter we have to deduct the initial time from the final time.

If Galileo had used a modern pulse-rate meter and measured his pulse continually through a 24-hour period he would have noted both short-term fluctuations and a long-term cycle. The pulse varies with demands placed on the heart. So if we run up a flight of stairs our heartbeat increases as the heart pumps blood more quickly to meet the demands caused by the exertion. But allowing for fluctuations caused by short-term physical or emotional demands, over a 24-hour period the rate itself shows a regular circadian rhythm. So the normal pulse at, say, noon has a different period from that at, say, 3.00 a.m., even when lying down and resting at both times. Just as a stopped clock is right twice a day, so the pulse is the same twice a day, once when on the rising part of the curve and once on the falling.

There is a similar circadian rhythm of body temperature, which is why it can be hard to decide whether someone has a 'temperature' or not – it depends on the time of day when it is taken, the shape of the individual's normal cycle and their immediate history. Our body temperature is seldom at 37°C although we try hard to keep it so. Human body temperature control is an example of negative feedback. Thermoreceptors in the skin detect the temperature of the external environment. This sensory information is relayed to the hypothalamus in the brain, which in turn transmits nerve pulses for corrective mechanisms to occur. When we are hot we sweat, which cools us down as the latent heat of evaporation is carried away to the external environment. This is compounded when the blood vessels close to the skin surface become more dilated, which provides a larger surface area for heat to be lost to the external environment from the blood vessels carrying overheated blood. Vasoconstriction is the opposite of this and occurs when the temperature in an organism drops. The blood vessels become constricted so that minimal heat loss occurs. When we are cold, the hairs on our bodies 'stand on end' and trap a layer of air between the hair and the skin. This insulation of warmer air next to the skin reduces heat loss. Other mechanisms are involved, such as a decrease in metabolic rate and shivering when temperatures drop.

At any moment, body temperature is skittering about a set point as these negative feedback mechanisms attempt to maintain the set point. But the set point itself varies over a 24-hour period, falling in the night and rising during the day (Figure 4.1).

Homeostasis is the name given to the organism's attempt to maintain a constant internal environment. It is one of the few big integrative ideas in biology, and some would say it is second only to evolution by natural selection. Just as evolution has outgrown its biological beginnings and become, in Daniel Dennett's phrase, a 'universal acid', so too has homeostasis become a guiding principle of analysis in many disciplines. The functionalist school in sociology, led by Talcott Parsons, believed a society homeostatically maintains its stability despite competing political, economic and cultural factors. Economists saw in homeostasis the inherent feedback in the law of supply and demand and the general

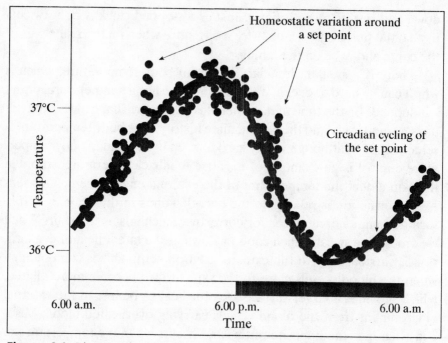

Figure 4.1. Diagram depicting the homeostatic and circadian variation in human body temperature. Body temperature constantly fluctuates at any moment as negative feedbacks (homeostasis) attempt to maintain the temperature at a set point. The set point itself cycles over a 24-hour period.

tendency of markets to find equilibrium levels. An easy way to make a living in the business world is to turn up at conferences and speak about the psychological theories of Abraham Maslow, whose hierarchy of human needs is based on a homeostatic process leading to what he termed self-actualisation.

Walter Cannon, a Harvard physiologist, coined the term (Cannon, 1947). Cannon was the first to show how to take X-rays of soft tissues in the intestine by giving bismuth and barium 'meals' to patients, but his interests changed to the control of physiological processes. In one experiment he measured the sugar in the urine of Harvard football players during a big game and he was fascinated by the way in which adrenaline controlled the release of sugars into the blood during the 'fight or flight' response. When insulin was discovered in 1922, Cannon was particularly struck by the finding that it had the reverse effect to adrenaline in that it caused sugar to be stored. This led him to the key idea of antagonistic agents, with one factor acting in one direction being opposed by a factor acting in the other.

The idea of constancy brought about by a dynamic interplay of antagonistic agents was a genuinely big idea and very simple in concept. The general principles are familiar to everyone who has ever driven on a motorway or set a central heating regulator. On the motorway we try to stay near a set point, which is usually the legal speed limit. The speedometer monitors the variable and our brain, which compares it with the set point, is the control centre. If the variable differs from the set point, the control centre uses effectors to reverse the change. One effector is the foot on the accelerator pedal, another is the foot on the brake.

The temperature for the central heating is set on the regulator, and sensors in the device continually monitor the immediate environment. When the temperature is below the set point, the boiler is fired up, and when it is too hot it switches off. The continual adjustment keeps the temperature at the required set level.

Blood sugar control is a homeostatic model of physiological regulation. The regulated function is monitored by an internal control centre that is sensitive to a relevant physiological parameter (in this case blood sugar, but it could be oxygen level or temperature or other factors). Sensors detect deviations from the optimal setting and stimulate behaviours or physiological responses that correct the deviation. Each time the variable being monitored departs from the set point there is a degree of overshoot and undershoot in the corrective action. Once the level is back to the set value on the regulator, the drive that stimulated the corrective behaviour is cancelled until the next deviation.

Provided the sensors are very sensitive and there is very little lag in the timing of the corrective action, it looks to the entire world as though there is a *fixité du milieu*. In reality, it is more like the motion of a swan in the water. Majestic serenity on the surface, constant and furious activity below.

Homeostasis, as it is commonly understood, is now viewed as somewhat more complex than simple negative feedback. Consider why an animal eats when it does. The quick answer is that it eats when it is hungry. If that were the case, it would behave as though there were a reservoir of calories and when this reservoir was depleted below a certain set point it would fill it up again. Likewise, it would sleep when it was tired and drink when it was thirsty as if there were a sleep reservoir and a fluid reservoir.

But animals do not behave like that. Syrian golden hamsters normally eat in many short bouts throughout the day and night. If hamsters are deprived of food for 12 hours each day and are allowed to eat freely from an unlimited food supply for the remaining 12 hours, they begin to lose weight gradually. The weight loss continues until the hamsters appear severely underfed. This is because the animals do not eat enough during the time when food is available to compensate for the food that they normally would have eaten during the daily period of food deprivation.

The animals do not cope well with significant periods during which food is not available. Hunger modulates the feeding pattern, but if a hungry hamster is allowed to feed at times different from the normal circadian range, the animal is unable to 'eat its fill'. The hamsters' feeding pattern is programmed by the biological clock and so they cannot eat for longer than they do even, when they are very hungry.

This temporal programming means that the animals eat not primarily because they have detected a lack of food, but as a mechanism to prevent any such deficits from occurring. Effectively, they do not wait until they are calorie deprived but they eat in a programmed manner to prevent themselves from becoming so. In this view, much of the behaviour of living organisms is anticipatory of regular and predictable daily environmental changes rather than reactive. It is based on feedforward rather than feedback mechanisms, be they negative or positive.

If we think again of the thermostatic central heating regulator, a more efficient system is one that is mechanically or electronically programmed to anticipate the predictable daily changes, for example that from Monday to Friday the home is vacant between 9.00 a.m. and 5.00 p.m. and consequently the boiler shuts down at 8.30 a.m. and fires up again at 4.30 p.m. The homeostatic mechanism in this case handles the short-range minute-by-minute fluctuations in the environment when the system is on.

A combined circadian/homeostatic mechanism explains the feeding behaviour in one study in which rats were food deprived for 14, 25 or 42 hours beginning early in the dark phase when rats normally start to eat (Bolles & Stokes, 1965). After 25 hours without food, the rats ate more than after 14 hours of deprivation beginning at the same time. But after 42 hours of deprivation, rats did not eat more than after 25 hours. In fact, at 42 hours they ate only about as much as after 14 hours of deprivation.

There is an underlying cycle so that the rat's hunger peaks in the night and falls during the day. Rats re-fed after 14 hours are being re-fed during the early light phase when they do not normally eat; those re-fed after 42 hours are also being re-fed during the light phase. By contrast, rats re-fed after 25 hours of deprivation are being re-fed during the night, when they normally eat. The greater food intake after 25 hours than after 14 hours of deprivation is predicted by both a homeostatic model and a clock model – by the homeostasis model on the basis that they are that much hungrier, by the clock model on the basis that they would normally be feeding at 25 hours anyway. However, the data at the 42-hour deprivation time are not explicable on a strictly homeostatic basis because the rats are even more food deprived, but are consistent with the predictions of a clock model; the endogenous timer should not be stimulating the animal to eat in the middle of the light phase.

The rat becomes hungry as a result of homeostatic processes, but just how hungry it becomes is a function of the phase of the circadian feeding cycle. In this framework, an organism's biological systems are programmed by a clock to meet the predictable and rhythmic daily variation in the two major environmental factors of light and temperature. This is the primary control system, and the homeostatic mechanisms are for fine tuning control in response to fluctuations in the external environment (such as moving from shade into light) or from within the organism (such as changing from a walk to a run).

Nicholas Mrosovsky cites the example of the environmental problems faced by a camel to explain the complementarity between clock programming and homeostatic control (Mrosovsky, 1990). Camels live in the middle of the desert. They are too big to burrow into the sand to seek cooler spots in the heat of the day and there is not much in the way of shade. The only other way they can keep cool is through evaporative cooling – sweating from the skin or from the tongue. But that means water loss, and the one thing a camel does not want to do is give up precious water.

The camel has a conflict. It can give up water to cool down, but then it dehydrates. There is a struggle between thermoregulation and water balance. It solves it with a neat trick. The camel's temperature rises during the day to levels that would be dangerous in most mammals. The keener the animal is to save water or the more dehydrated it is, the higher its temperature goes, peaking at perhaps 41°C. That is a raging

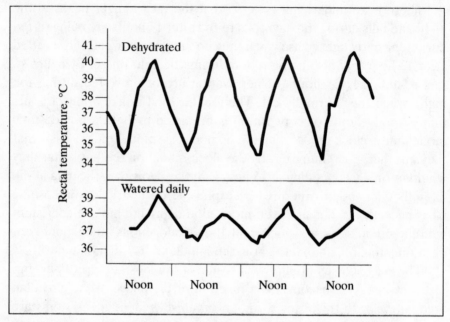

Figure 4.2. Diurnal changes in rectal temperature of a camel, showing nocturnal hypothermia in dehydrated animals. From Mrosovsky (1990).

fever and a life-threatening temperature for humans, but not for the camel.

At night, the dehydrated camel allows its body temperature to fall as low as 34.2°C, which is very cool by human standards. The advantage to the camel is that the following day it takes longer to heat up to high levels. The animal uses nocturnal hypothermia to anticipate and protect itself from the next day's heat in an example of a feedforward mechanism (Figure 4.2).

The camel's internal environment is not invariably fixed. It can alter the set point to suit the environmental conditions and its own level of hydration. The car driver analogy would be changing from a motorway to an urban area and aiming to keep the speed at say 30 miles per hour instead of 70. But even though the camel has flexible temperature set points, there is an optimal range and the overall daily temperatures are still under circadian control. The period of the camel's diurnal temperature rhythm is constant but the amplitude can change (Schmidt-Nielsen *et al.*, 1957).

Biochemical systems are dynamic. There is constant fluctuation even in a constant environment. Every living creature is a frenzy of feedback. If we could drill down into a cell we would see atoms and molecules careering around and the main impression would be that of constant motion, constant arranging and rearranging. It is like riding a bicycle, perfectly possible as long as it is moving.

Overlying this constant dynamism is a regular temporal dimension. This continual, monotonic rhythm determines an organism's spontaneous behaviours and responsiveness to external events. The variation in feeding behaviour described earlier is not random or capricious but is part of a predictable temporal structure imposed on organisms by their evolution in a physical and biotic environment that is itself rigidly structured in time.

Jürgen Aschoff, another circadian pioneer who headed the Max Planck Institute for Behavioural Physiologie, explained this linkage between homeostasis and circadian rhythms 40 years ago (Aschoff, 1964):

> *Homeostasis is a shielding against the environment, one might say, a turning away from it. For a long time, this phenomenon has been taken as the prime objective for an overall organisation in physiology; and it evidently has great survival value. But there is another general possibility in coping with the varying situations in the environment; it is, instead of shielding, 'to turn toward it'; instead of keeping the 'milieu interne' stable, to establish a mirror of the changing outside world in the internal organisation. This has one clear prerequisite; the events in the environment must be predictable, which of course is the case when they change periodically.*

Like the judoka who is trained to turn an opponent's momentum to her advantage, organisms mould their lives around the natural rhythms of their environment rather than try and work against it. If the animal or plant in the wild has the means for predicting regular environmental changes, then it can behave in such a way as to exploit the conditions. So a bird that is aware that dawn is approaching is able to be physiologically prepared for first light in advance. It has to anticipate dawn so that its biochemistry is 'ramped up' to meet the demands of the day. In this way, the early bird that catches the worm has the better temporal programme; 24-hour rhythmicity provides a vital temporal programme that enables an organism to anticipate the exigencies of the day.

5

THE SEARCH FOR THE CLOCK

Everyone compares himself to something more or less majestic in his own sphere, to Archimedes, Michelangelo, Galileo, Descartes, and so on. Louis XIV compared himself to the sun. I am much more humble. I compare myself to a scavenger; with my hook in my hand and my pack on my back I go about the domain of science picking up what I can find.

STATEMENT BY THE FRENCH PHYSIOLOGIST MAGENDIE, WHICH HE
ADDRESSED TO HIS STUDENT CLAUDE BERNARD, SEEN ON A PLAQUE IN
THE LATE PROFESSOR CURT RICHTER'S LABORATORY

Scientists, particularly serious scientists, are not like the rest of us. Some are labelled delightfully eccentric when in truth they are barking mad. Others are called dedicated when compulsive might be a more appropriate word. When James Watson implied, in his book about discovering the double-helical structure of DNA, that scientists may even be interested in sex as pleasure rather than as an object of study, he was greeted with opprobrium in the scientific community – though it has to be said that they had many other more legitimate reasons to feel that way about his book.

Yet odd they are. It goes with the territory because to succeed, a scientist not only has to pick the field of study with care but also has to be more than a touch obsessive. Curtis Richter was a great scientist. But after he finally, totally, completely retired, researchers who went through his lab records had to put on protective clothing, face masks and filtered breathing devices. Richter had worked in the same laboratory at Johns Hopkins in Baltimore for 63 years. Throughout that time, according to Tim Moran, now a professor in the department of psychiatry at Johns Hopkins and who knew Richter well, 'Richter never wanted work in the laboratory interrupted for painting or general maintenance. Consequently, by the time the laboratory was closed in 1988, paint that dated back to the early 1920s was peeling from the ceilings and walls. All the records had become highly contaminated with lead dust' (Moran & Schulkin, 2000). No one has ventured to suggest what it might have done to the experimental animals.

Every branch of science has its unsung heroes. Galileo gets the credit for the pendulum, whereas strong claims can be made for prior discovery by Mersenne; Alfred Wallace lost out on evolution to Darwin; Crick, Watson and Wilkins shared the Nobel Prize for their discovery of the structure of DNA while Rosalind Franklin, whose painstaking work provided the X-ray photographs that revealed the double helix, sadly died too soon and got nothing.

Richter is hardly unsung as a pioneering physiologist, but it is not often recorded that during his monumental life – he died in 1989 aged 95 – he carried out some of the earliest studies on rhythmic behaviour. There are many claims to the parenthood of animal chronobiology, but his are stronger than most.

He came to science fairly late. Born in Denver in 1894 to German immigrant parents, he studied in Germany before the advent of the First World War suggested that he might be better off back in the USA. After Army service he became interested in biology and presented himself to John Broadus Watson, then the professor of psychology at Johns Hopkins (Richter, 1985).

Watson took Richter on as a doctoral student and then left him completely to his own devices. Current PhD students can only stand and marvel, but that is how it was done back then. With no stimuli to which he could respond, Richter had not a clue what to do. After a week he found that a cage with 12 rats had been placed in his room. Not sure whether Watson had sent them to give him ideas or simply for company, he decided to study their spontaneous activity. Never having seen a rat close-up before, he started from scratch. He built special cages so that he could continuously monitor the animals' weight, food and water intake, and movement. He took to the work like the proverbial duck to water, or, as he put it when he talked about Watson, 'he released my gene for freedom of research'.

There is more than a touch of irony in that Richter worked with animals from whom he deliberately withheld stimuli, in the laboratory of the man who held that without stimulus there was no response. Watson was a founder of behaviourism and it was his conviction that a few things are instinctive, but virtually anything can be taught. He believed emphatically that nearly everything in an animal's behaviour was due to nurture and hardly anything to nature.

By blacking out the windows, soundproofing the lab room and ensuring that extraneous factors and external stimuli were absent, Richter was able to log each rat's activities when it was left to its own devices. With his special cages he could monitor a rat's activity, including not only when it moved but when it fed and drank, how much food and water it took, when it defecated, and so on.

By the time Richter had his doctorate in 1922, Watson had been fired for having an affair with one of his research assistants (the departmental chair who had hired Watson had himself been sacked some years earlier when caught in a brothel, so maybe it was something in the Baltimore air). Richter was put in charge of his lab and there he stayed for 63 years.

Richter's early work showed that the spontaneous activity in the rat was rhythmic. He was not the first to study cyclical behaviour in mammals. Moore-Ede has described how the British physician William Ogle wrote in 1866 about the daily body temperature in humans (Moore-Ede *et al.*, 1982):

> *There is a rise in the early mornings while we are still asleep, and a fall in the evening while we are still awake, which cannot be explained by reference to any of the hitherto mentioned influences. They are not due to variations in light; they are probably produced by periodic variations in the activity of the organic functions.*

In the early 1900s, Simpson and Galbraith measured the rectal temperatures of five monkeys (manually, it should be recorded) every two hours for 60 days. They showed that there was a persistent body temperature rhythm of about 24 hours in both constant darkness and constant light (Simpson & Galbraith, 1905).

But according to Paul Rozin (Rozin, 1976), Richter's work

was probably the first systematic study of the range of biological rhythms and their determinants in mammals and possibly in vertebrates. Already in the 1921 studies Richter noted that in shifting from a lighting schedule of 12 hour light, 12 hour dark to continuous darkness (L:D 12:12 > DD), the pattern of high activity in the previously dark 12 hours remained for some 12 days, strongly suggesting endogenous origins.

His early study was remarkable not just for his description of rhythmic behaviours but also for his innovation: the running wheel that is the staple of the activity studies of small rodents was one of his inventions. His work on the importance of spontaneous activity and his demonstration that it was internally generated foreshadowed in many ways the big issue of neuroscience in the 1950s: to what extent is brain activity an intrinsic organ function, independent of stimuli from the outside world?

Why an animal does one thing at a particular time and not another is a deep issue in ethology and also, by extension, in human behaviour. Cat owners know that their pets have well-defined routines. At given times they eat; at others they walk about, stretch and go outdoors. Their sleep, unlike a human's, is not consolidated into one session but they cat-nap, sleeping in selected places at different times.

Watson and later behaviourists (after being forced out of academia, Watson made a fortune in advertising) ascribed the cat's behaviour to a sequence of stimulus–response chains. The idea was that the environment provides a string of stimuli that trigger a matching stream of adaptive responses. The problem with this mechanical approach is that cats do not respond in exactly the same way to the same stimulus. Nor do we. It depends on the context and what has happened before. Economists talk of the law of diminishing returns when they explain how an individual becomes satiated and so will spend less on, say, another strawberry when she has already eaten half a kilo. The twentieth strawberry does not evoke the same appetite response as the first.

The opposite also occurs. The marginal value of food can increase. It took them a while to get over their inhibitions, but eventually the rugby players whose plane crashed in the Andes were hungry enough to eat their dead team-mates.

The behaviourist explanation was gradually abandoned and motivation became the dominant model. The idea was that there was a range of homeostatic drives or 'motivations' that determined behaviour. So an animal became increasingly hungry as its calorie reserve dwindled, and at a set point it would feed. When it had filled the calorie reservoir, the hunger drive would diminish and be replaced by another, perhaps thirst as the drive and drinking as the behaviour.

In effect, there was a rotating repertoire of drives and their concomitant behaviours, each drive taking its place in turn at the top of the hierarchy. But this still left unanswered a very big question. It did not explain Richter's findings in the rat that there was regularity to a given behaviour. Why was it that particular drives regularly appeared at certain times of day and not at others?

The answer comes from close observation of the rat's behaviour. A nocturnal animal is active at night and sleeps during the day. It uses up far more energy during the night than it does during the day, so all the physiological parameters are higher at night than during the day. It does certain things during the night and certain things during the day because that is the evolutionary strategy that has led to success. This differential metabolic rate is encoded in the animal's genome. It lives this way because that is the way it is adapted to a predictable environment. And this is controlled by a clock.

However, the environment is not totally deterministic. Although it is broadly predictable, with night following day and winter following autumn, there are fluctuations in the local environment. The temperature may be a little lower and so the animal will have to work a little harder. The homeostatic mechanism takes care of these fluctuations, but it is a fine tuner rather than the main controller.

Richter's early work led to the realisation of the clock's central role in determining when an animal, plant or amoeba does what it does. But the realisation of the universal biological significance of this only came after certain prosaic questions had been answered. Such as, if there is a clock, where is it? Richter's seminal contribution was in helping to find the location of the mammalian clock. He conducted what can only be described as a heroic series of experiments, first to confirm that it was the brain that housed the clock and secondly, to try to establish where exactly it was in the brain (Richter, 1967).

He used 'almost every conceivable kind of metabolic, endocrino-

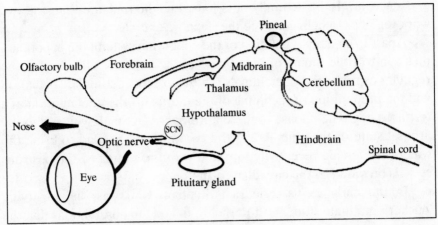

Figure 5.1. Diagram of the rat brain sectioned down the mid-line, showing the cut surface of the left-hand half of the brain. The major regions of the brain are shown. Lesions of the frontal part of the hypothalamus, which destroy the suprachiasmatic nuclei (SCN), abolish rhythmic circadian behaviour. Light from the eyes reaches the SCN via the optic nerves, which contain a dedicated projection to the SCN called the retinohypothalamic tract. Note that the thalamus and hypothalamus are regions of the forebrain, not the midbrain, and that the midbrain structures are relatively reduced in mammals. The total length of the rat brain is about 2.5 cm.

logic and neurologic interference' (Moore-Ede *et al.*, 1982), to see their effect on the free-running activity rhythm of blind rats. Richter removed adrenals, gonads, pituitary, thyroid, pineal, pancreas. He gave his rats electroshock therapy, induced convulsions and prolonged anaesthesia, and with suitable irony even got them blind drunk.

Still they were rhythmic. It was only when he turned to the brain that he was able to produce arrhythmia. He made over 200 studies investigating what happened if he damaged the rat brain and measured the effect of the lesion on the rat's locomotion, feeding and drinking behaviours. In 1967, Richter reported that lesions in the front part of the hypothalamus of the brain eliminated multiple behaviour rhythms (Figure 5.1). But he could not get any closer.

The hypothalamus is at the base of the brain. In humans it is about the size of an almond and in rats smaller than a cardamom seed. Its main function is to regulate such factors as blood pressure, body temperature, fluid and electrolyte balance, the metabolism of fats and carbohydrates, and sugar levels. Structurally, it is joined to the thalamus and the two work together to help generate the sleep/wake cycle.

Apart from daily regulation of the endocrine system, the hypothalamus controls the timing of the release of the hormones involved in the regular oestrous cycle of mammals. These hormones stimulate ovulation and the preparation of the lining of the uterus for the implantation of a fertilised egg. The release of these hormones is intimately linked to the arousal state of the animal, and the coordination needed to bring the window of female receptivity into line with the timing of the oestrous cycle is provided by the circadian oscillator.

In the early 1970s, Friedrich Stephan, who was one of Irving Zucker's graduate students at Berkeley, linked the observations that the oestrous cycle was dependent on the circadian control of the timing of the hormone release and a small part of the brain in the anterior hypothalamus. By careful lesioning in the frontal part of the hypothalamus he identified a small paired cluster of cells known as the suprachiasmatic nuclei (SCN). Loss of the SCN abolished the oestrous cycle and rhythms in behaviour such as drinking and locomotion (Figure 5.1) (Stephan & Zucker, 1972).

At the same time that Stephan and Zucker were performing lesions, Robert Moore, then at the University of Chicago, was using a different approach to find the clock. Many studies had shown that there was an intimate relationship between the clock and light. Depending on the point of the circadian cycle at which they were delivered, light pulses could either advance or retard the clock. This process of setting the clock was known as entrainment.

Moore's approach was to follow the light beam as it came in through the eye and to see where it went. There was one well-known multi-step pathway that took light via the optic nerve to the visual processing regions of the rat forebrain. It was known, however, that if the optic nerves were cut before the optic chiasm (where the optic nerves from the two eyes cross over at the base of the brain) then light/dark cycles no longer entrained the biological clock; but cutting just after the optic chiasm left the ability of light/dark cycles to entrain the clock unaffected. This suggested there was a distinct pathway separate from the visual projection that was responsible for transmitting the signals for setting the clock (Moore & Eichler, 1972).

In 1972, he and his co-workers injected radioactive amino acids into rats' eyes. They followed the tracer molecules as they travelled along from the eye, through the optic nerves, and along a projection into the SCN. The newly found pathway from the eye to the SCN was called the retinohypothalamic tract.

When Moore lesioned the SCN, there was a loss of the circadian rhythm of the hormone corticosterone. He had shown that the SCN was involved at the very least in the circadian timing of neurosecretion. Between them, Moore, Stephan and Zucker had established that removing the whole of the SCN (but not just a part of it) destroyed both behavioural and endocrine circadian rhythms. Just 20,000 or so cells seemed to be responsible for controlling the timing of a mammal's endogenous rhythms.

What is true of rats is probably also true of humans. Close your eyes and imagine there is a line inside your head from the bridge of your nose to the base of your skull. Now imagine another line drawn across your skull about two centimetres in from the eyes. Focus inwardly on the spot where they intersect. Buddhists name this spot the 'third eye' as it is supposed to open in enlightenment. They believe that if you focus on this spot and empty yourself of all other thoughts, this inward contemplation should bring calmness into your life. The SCN is at the intersection of the imaginary lines and in many ways the Buddhists may be on to something. Nobody has suggested carrying out lesioning experiments on living humans, but this tiny group of cells, less than a third of a cubic millimetre in volume, has been called the 'mind's clock'. Other workers have shown in birds and reptiles that there is similar arrhythmicity when specific structures that can be thought of as analogous to the SCN in mammals are destroyed.

These studies all raised an obvious question. Was the SCN the clock itself in mammals or was it merely a link in a chain? The early thinking was that it might be merely a relay station. There was undoubtedly an association between the SCN and rhythmicity, but the SCN might simply be part of a clock system and the true oscillator might be located elsewhere. The SCN might be necessary to maintain circadian rhythmicity of certain functions, but was it also sufficient for a full explanation? It was the potential anatomical link between the external and internal environments, and that led to a neat model. The eye sensed light and a dedicated pathway could capture this light and deliver a sensory input to an internal structure, namely the SCN, that was in turn connected to the neural and endocrine systems. But was the SCN itself the clock that linked this external information with the timed release of hormones?

Finding the 20,000 cells in the SCN out of the several billion in the rat brain was tough enough. Proving that this structure was also the seat of the biological clock in mammals was even tougher. The intellectual challenge was to show both that the SCN itself generated an oscillation when isolated from all inputs and that it was the timing driver for other rhythmic functions.

It took nearly 20 years to establish incontrovertibly that the SCN in mammals was a clock that regulated several functions with a period of about 24 hours. Most of the work was done in laboratories in the USA and Japan. It is an excellent example of the way in which biological science works in trying to distinguish between associations and causal chains and is worth recounting in some detail. There is a caveat. The research followed different paths, and hindsight provides the effort with a more coherent structure than it warranted. The point about biology is that it is interdisciplinary, and a major effort now goes into synthesising observations and experiments from widely disparate sources to come up with a picture of biological reality. It is by its nature a messy business.

To recap, by the mid-1970s it was well established that destroying the SCN affected the timing of various behaviours in mammals but did not abolish the behaviours themselves. So rats that had their SCNs destroyed would still sleep, eat and drink but there would be no 24-hour pattern to their activities; they were arrhythmic.

The next step was to find out what happened when the SCN was isolated within the living animal. This was easier said than done because the cells of the SCN are located very close to several major nuclei and tracts in the brain. This made it all but impossible to restrict lesions to the SCN without damaging any of the fibres or neurones surrounding it.

In a very elegant study in the late 1970s, Shin-Ichi Inouye and Hiroshi Kawamura, working at the Institute of Life Sciences in Tokyo, demonstrated that the SCN neurones generate a circadian rhythm of electrical activity. A very fine electrode was inserted into the brains of rats. Each electrode was only 200 μm wide at the tip (a human hair is about 100 μm wide) and measured the electrical activity of multiple neurones at the same time. Inouye and Kawamura compared the responses of neurones inside the SCN with that of neurones close to but still outside the SCN (Inouye & Kawamura, 1979).

They found that the neurones within the SCN showed a circadian pattern of rhythmic electrical activity that was high in the day and low at night. The neurones outside the SCN also showed a circadian rhythmicity but they were 180° out of phase; in other words they peaked in the night and were at their lowest during the day.

When Inouye and Kawamura isolated the SCN by cutting around it with a rotating knife, severing its neural connections to the rest of the brain, the pattern continued within the SCN itself but rhythmicity was abolished in the nearby neurones that were outside the cut. This suggested that not only was there an endogenous rhythm produced within the SCN, but also that electrical signals carried by the neural connections leading away from it – the efferents – provided timing information to the other brain regions.

This was an important finding but it was still possible that there was another factor outside the SCN that was the real oscillator. Maybe there was a hormone that was produced rhythmically elsewhere that diffused into the anterior hypothalamus and was driving a rhythmic signal in the SCN. In other words, it was thought that the SCN might be a critical relay station from the 'true' clock. Another suggestion that used to irritate intensely the ultra-polite Kawamura, because he took it as a slur on his careful studies, was that perhaps not all the neural connections going to the SCN had been cut.

The SCN had to be dissected out of an animal and examined in complete isolation to see whether it was truly oscillating. Fortunately, tissue from the central nervous system survives for about six minutes without damage, which was enough time to slice out a region of interest and put it into a tissue-support system that would keep it going with nutrients and oxygen. Measuring the electrical activity showed that the SCN generates a stable rhythm with a period of about 24 hours, and was a self-sustaining oscillator.

At the University of Massachusetts, Bill Schwartz had been looking for some time at the metabolism of the SCN to study its function, based on the seemingly obvious observation that working tissues consume energy. This is simple enough for homogeneous organs such as the heart and kidney, which perform repetitive and easily measured work. The brain is a different proposition. It needs a continuous supply of glucose for its energy needs, but establishing the relationship between brainwork and energy distribution is a whole new can of worms. The brain is heterogeneous, and different parts do very different things. Deciding what is meant by work is neither easily defined nor easily measured.

Schwartz injected radioactive 2-deoxyglucose (2DG), which is very similar to glucose, into a rat's vein. (2DG is carried to the brain, but unlike glucose it cannot be metabolised any further to provide energy, so the 2DG molecules are effectively 'fixed' in the brain cells.) About 45 minutes later, the animals were killed and thin slices of the rat brain were placed on X-ray films. The developed films clearly showed the location of the radioactively labelled 2DG, and the intensity of the image indicated the amounts of the molecule present in the tissue slice (Schwartz *et al.*, 1980).

Schwartz found that the SCN is metabolically active during the light phase of a 12 hours light/12 hours dark cycle and relatively inactive during the dark phase. Significantly, no other brain region exhibits such a dramatic rhythm. Subsequent work on a wide range of rodents has shown that the rhythm of glucose utilisation in the SCN is indeed circadian.

This was another piece of evidence that the SCN was the generator of a circadian rhythm. In the middle of the 1980s, workers in several laboratories showed a direct effect when they restored circadian activity rhythms in SCN-lesioned adult rats and hamsters by transplanting foetal SCN. Animals whose SCN had been destroyed slept and woke for a few minutes at a time without a discernible pattern, but regained their activity rhythms once they received the transplant, confirming the SCN's importance as a critical part of the circadian system.

But this was still not enough. Was it only SCN cells that were being transplanted? Could there not have been some other cells attached? And it could still have been the case that the SCN was simply part of a chain rather than the whole story. The transplant might have been producing some 'stimulating factor' that activate the dormant oscillator of the host. It could still have been an important but nevertheless secondary structure and not the main driver.

It may seem that the sceptics were going overboard, but they had a point. The evidence was accumulating but the killer experiment had still to be done. The chance discovery of a mutant golden hamster in Michael Menaker's laboratory at the University of Oregon (he subsequently moved to Virginia) provided the means for just such a test (Ralph & Menaker, 1988).

70

Thousands of golden hamsters are found in biology departments all over the world, yet they all derive from one original breeding pair. The advantage of this inbred strain is that variation between animals is reduced. Variation is a huge problem for experimental biologists, and it is this that makes mutants so important. They provide biologists with a difference that is large enough to be differentiated from random variation.

Physicists have it easy. At the elemental level all the particles in a set are the same. Steven Weinberg, the Nobel laureate, once said, 'if you have seen one electron you have seen them all' (Weinberg, 1998). Biologists have to continually ask whether they are seeing a real effect or something within the normal range of variation. For instance, if the daily rhythms of heart rate and body temperature of, say, three golden hamsters are measured hourly for six consecutive days, at any given point no two hamsters will have identical heart rates. They vary between themselves by plus or minus 20 beats a minute. Similarly, the body temperature varies by ± 0.1°C at any given point. So experiments using one of the three hamsters as a control have to allow for this normal variance.

Not only is there variation between individuals; there is also considerable variation within an individual. Both the hamster heart-beat rate and the body temperature oscillate on a 24-hour period. The heart rate goes as low as 360 beats per minute in the day to a high of 480 beats per minute when the hamster is active at night. The body temperature drops to a low of 36.3°C in the day and rises to a high of 38.0°C at night. Similarly, human body temperature varies daily from a low of approximately 36.5°C to a high of about 37.4°C. At about 5.00 p.m. the temperature reading of 37.0°C not only is not a fever but is actually below the normal expected reading of 37.4°C. In fact, the traditional definition of fever as a temperature above 37°C applies only at certain times of the day. There is a strong argument that every experimental or clinical observation on living organisms should be time-stamped!

The amount of variation in a population is known as a Gaussian distribution or more familiarly as a bell-shaped curve (Figure 5.2). We understand this intuitively from simple observation. A group of adults of the same age will have different heights, weights and body shape. Some of the variation between them will be environmental and due perhaps to what they eat and how much they exercise. Some of the variation will be down to genetic factors. The interplay between the genetic and environmental

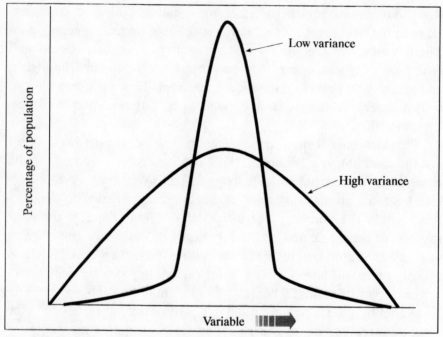

Figure 5.2. The Gaussian distribution of an unnamed variable in two populations. One population shows low variance, with most individuals in the population clustered around a narrow range or mean, whereas the second population shows a high degree of variance.

factors, particularly during development, results in the individual organism, known as the phenotype.

The more similar the people are and the less their variation, then the narrower the Gaussian distribution is. The more dissimilar, then the flatter the curve. If every human being were exactly the same height at the same age, give or take a millimetre or two, then we could say that the variation was very low and that there was very tight genetic control. Given how different we are, it is reasonable to say that there is considerable variation and that although there is some genetic determinism – children do at least resemble their parents – our height phenotype is not tightly clustered.

Steven Rose has explained in *Lifelines* (Rose, 1998) that the phenotype is used to refer to any or all of the observable or measurable features of an organism, from the presence of a particular enzyme to hair colour or body feature, or even a piece of characteristic behaviour such

as the gait of a walker. Circadian rhythms are just such pieces of characteristic behaviour. Within a species, the circadian rhythm between different individuals will vary even if the individuals themselves are genetically identical. For example, *Gonyaulax* is an alga that reproduces mitotically, that is it simply divides in two. It is basically a cloned colony and every individual has the same genetic make-up. Even in this case of identical organisms, the period of the circadian rhythm of these creatures shows some variation.

At this point, the sceptical inquirer would rightly ask whether this Gaussian distribution in *Gonyaulax* is real or nothing more than an artefact of the assay process? In other words, is the variation in the assay procedure greater than the variation in the circadian rhythm? In the case of *Gonyaulax*, measurement of the circadian rhythm is very accurate, so the distribution is probably not an artefact. This is precisely the value of circadian rhythms in studying behaviour. They provide a clearcut model for the genetic control of easily measured objective behaviour.

Martin Ralph, who was working on his PhD in Mike Menaker's laboratory, was assaying the rhythms in a new batch of hamsters by monitoring their activity on running wheels. Golden hamsters usually have free-running circadian rhythms close to 24.1 hours when kept in constant darkness. A few hamsters in a population have rhythms not much above 23 hours and, at the other end of the distribution curve, a few show rhythms of about 24.5 hours, so it is a fairly tight distribution pattern, which is true of most circadian rhythms. In many species, the interindividual variance in the rhythm is less than 0.5 per cent.

Ralph found that one of the hamsters in the new batch had a circadian activity period of 22 hours, which might not seem that much but in terms of hamster circadian rhythms it was far shorter than anything seen before. It was not so much a mutant as a monster! Fortunately the mutant had gone to a lab working on circadian rhythms. If it had gone to another lab, the circadian variation would not have been noticed, proving yet again the importance of serendipity in science. This hamster was named the *tau* mutation, and Ralph and Menaker realised that they had the tool to see whether the SCN was unambiguously the key oscillator. If the SCN was taken from the mutant and transplanted, would the recipient show a circadian period of the donor or of the recipient? This was a testable proposition. If the period was that of the donor, this would indicate that the transplant reflects the genotype of the donor and not that of the host; ergo the SCN must contain the oscillator.

Transplantation works best with foetal tissue, so Ralph's first task to breed his mutant hamster to obtain foetuses. When the *tau* mutant was bred with the normal hamsters, half the offspring had a rhythm of about 22 hours and the other half the normal 24.1 hours. The offspring with a 22-hour period were then back-crossed with another 22-hour *tau* mutant and three clear populations emerged.

Population 1 had an average or mean circadian period of 20.2 hours. In this case the offspring inherited a *tau* mutant gene (more properly known as an allele) from each of its parents. These animals were therefore homozygous for the *tau* gene. Population 2 had an average circadian period of 22 hours and in this case animals had received a *tau* allele and a normal allele from its parents, and so were what the geneticists call heterozygotes. Population 3 had the normal circadian period of 24.1 hours, receiving the normal allele from each of the parents, and were homozygous for the normal gene. The variation of the circadian period about the mean for each of the three populations was low, and there was no overlap of the three distribution curves (Ralph & Menaker, 1988).

Ralph and Menaker, and their colleagues Russell Foster and Fred Davis, then transplanted the SCN from foetal *tau* mutant hamsters into normal-type hamsters that had become arrhythmic after their SCNs were destroyed. In every case the restored rhythm showed a period close to 20.2 hours. Likewise, when the transplantation was done the other way round, from normal to mutant, again it was the period of the donor that was restored. This specificity of the period was the unambiguous proof needed to satisfy even the severest sceptic. The hypothesis that the SCN contained a circadian oscillator was correct (Ralph *et al.*, 1990).

Soon after the *tau* mutant work, other researchers found that even individual cells in the SCN show a circadian rhythm. A SCN was put into a culture solution and the cells were then dissociated from each other while supported on a multi-electrode array. This smart piece of electronics is a grid of tiny electrodes over which the SCN neurones spread. SCN neurones fire off at a frequency between 0.3 and 9.9 Hz. That means that at some times they hardly fire at all, while at the other extreme they are firing every 20 milliseconds or less. Steven Reppert and his team from

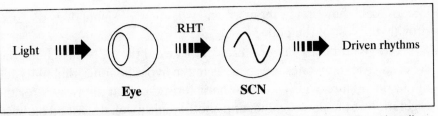

Figure 5.3. A simplistic view of the organisation of the mammalian circadian system. Light, detected by the eye and conveyed to the SCN via the retinohypothalamic tract (RHT), entrains a circadian clock within the SCN that in turn drives rhythmic physiology and behaviour.

Harvard found that the overall firing rate of individual rat SCN neurones showed a marked circadian rhythmicity averaging 24.35 hours, with peak rates of firing occurring between about 8 and 16 hours into a 24-hour cycle (Welsh *et al.*, 1995). So not only was the SCN the anatomical site of the oscillator but it was composed of cells that themselves oscillated, and that somehow coordinated their individual firing to give an overall rhythm of just over 24 hours.

The simple model of mammalian circadian organisation, outlined in Figure 5.3, held that the circadian clock in the SCN produces a rhythmic output that drives circadian rhythms of activity: drinking, feeding, sleeping, and so on. The oscillator rhythm is locked on to local time by the daily entrainment of dawn, dusk or both.

The model shows one master clock controlling all the rhythmic activity. The difficulty with this concept is that the master clock would have to be constantly sending out signals to the different parts of the body at different times. The liver would need timing signals so that it could prepare itself in advance of food intake; the eyes would need their timing signal telling them that dusk was approaching and to shift to rod vision; and the pineal gland would need to be informed that night would soon be falling and that the production of melatonin should start. The SCN would be like the control tower at a busy airport that has to coordinate the arrival and take-off of many different planes throughout the day and night, continually sending messages to different parts of the animal to coordinate its rhythmic timetable. Like whirlwind tourists on a trip round Europe, the SCN would be metaphorically saying to itself, 'If it is 9 o'clock in the morning it must be the kidneys and if it is 11 it must be the liver.' This model raised two critical questions: could the circadian system really function in this way, and what type of signal does the SCN provide the rest of the body?

A large number of neural projections from the SCN have been discovered. The main ones are to areas in the hypothalamus, thalamus and midbrain (Figure 5.1). The best-characterised output pathway is for the regulation of the pineal gland, and another fairly well described pathway is for the rhythmic control of corticosterone. Neurones in the SCN send direct and indirect projections to the neurosecretory corticotrophin-releasing factor (CRF) neurones in the hypothalamus. These CRF neurones regulate the release of adrenocorticotrophic hormone (ACTH) from the anterior lobe of the pituitary gland, and the rhythmic release of ACTH from the pituitary drives the release of corticosterone from the adrenal glands. Corticosterone has a key role in an animal's metabolism because it helps regulate the conversion of amino acids into carbohydrates and glycogen by the liver, and helps stimulate glycogen formation in the tissues.

Transplantation studies, however, suggested that neural connections were not the only contacts made by the SCN with the rest of the organism. It had been assumed that the recovery of circadian rhythmicity after SCN transplantation was due to some neural connection that originated from the SCN, but a transplanted SCN often makes fairly feeble neural connections with the host brain. Although a transplanted SCN restores rhythmicity in rodents in which the SCN has been removed, in no case has the transplanted SCN ever been entrained by light. The host animals show free-running rhythms, even under a light/dark cycle. This suggests a very limited capacity to re-establish neural connections with the host. Furthermore, the restoration of rhythmicity can be very rapid, too short a time for even minor connections from the SCN to the brain to be established. Even an SCN transplanted into the ventricles (the chambers of the brain in which cerebrospinal fluid circulates) could restore rhythmicity. It was difficult to see how the SCN could send out projections from this site to reach the hypothalamus or thalamus to control behaviour within a few days. Might the signal from the SCN be chemical as well as neural?

In 1996, Rae Silver's group from Barnard College in New York did not transplant the SCN into the brain directly but first placed the SCN in a semi-permeable polymer capsule (Silver *et al.*, 1996). The small pore-size of the capsule would not allow the passage of neural fibres but allowed diffusible substances out of the SCN graft. Transplanted SCN

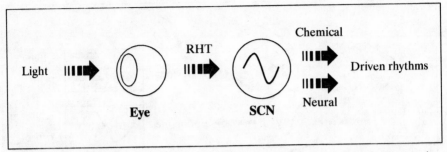

Figure 5.4. By 1996 the SCN was thought to drive rhythmic physiology and behaviour using both direct neural connections and the rhythmic release of unknown chemicals or 'humoral signals'.

contained within a capsule restored behavioural rhythmicity! So there is a 'diffusible chemical factor' from the SCN that helps organise wheel-running behaviour. The SCN talks to the rest of the body using both direct neural connections and unknown diffusible chemical substances (Figure 5.4).

But from the mid-1990s this view of the mammalian circadian system, with a 'master' clock in the SCN driving the circadian rhythms in the body, started to unravel. It was known that insects have multiple circadian oscillators. There is evidence of circadian rhythms in clock gene abundance in the wings, legs, oral regions and antennae of *Drosophila*. Jadwiga Giebultowicz of Oregon State University has found a circadian clock in the malpighian tubules in the fly's kidney (Giebultowicz & Hege, 1997). When she decapitated the flies the cycle did not damp out but persisted indefinitely, and in flies these peripheral clocks are entrained by local tissue photoreceptors. But mammals were thought to be very different from other animal phyla in regard to multiple clocks, although there had been hints that they might possess them. Richter's work in the 1920s had shown that when food was available for only a limited time each day, rats increased their locomotor activity two to four hours before the onset of food availability. This anticipatory behaviour has been found in other mammals and in birds and is accompanied by increases in body temperature, adrenal secretion of corticosterone, and activity of the digestive enzymes. Provided the food is made available at approximately 24-hour intervals, the animals can become entrained and show this anticipatory behaviour, even when they have had their SCNs destroyed.

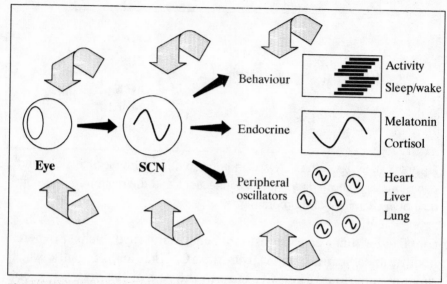

Figure 5.5. The current schematic view of the circadian system in mammals. The SCN provides signals that drive rhythmic outputs and entrain the rhythmicity of peripheral oscillators. The light/dark cycle is detected by the eyes, which in turn entrain the SCN. However, behavioural and physiological signals are likely to feed back at all levels along the circadian 'axis' to fine-tune the rhythmic output of the whole organism.

Although there had been these hints, there was general scepticism as to the universal presence of clocks in all peripheral tissues in mammals. In a surprising series of experiments, Ueli Schibler at the University of Geneva overturned the scepticism. He initially showed that a cell line of fibroblasts (cells found in connective tissues) that had been cultured for about 30 years could be induced to show a 24-hour pattern of gene expression! He was able to induce 24-hour cycles of expression by simply treating the cultured cells with serum. Somehow one of the many ingredients in the serum (or a combination) was capable of initiating a 24-hour oscillation of gene expression for a few cycles before the oscillation flattened and damped out. The rhythm could be induced once again with a further serum shock (Balsalobre *et al.*, 1998).

So these cells, which had been cultured in isolation in the lab for 30 years, and with no contact for all that time with an SCN, still retained the capacity to generate a circadian rhythm. We still do not know how the serum 'shocks' the fibroblasts into activity, but these and later experiments in other tissues such as the liver and heart led to the realisation that cells in the body seem to contain the machinery to drive circadian rhythms in gene expression.

The straightforward conclusion is that the vertebrate body is composed of billions of independent clocks. But there is a critical difference between a local clock and the SCN. Only transplanted SCN cells can restore rhythmicity to an SCN-lesioned animal. The transplantation of fibroblasts, or even cells from other regions of the brain, will not make an arrhythmic animal rhythmic. This suggests that the SCN regulates a temporal programme. A circadian organisation based upon multiple oscillators has been likened to the generation of 'standard time'. At the beginning of the nineteenth century there were 144 official times in North America. A town could reckon its local time from noon, when the sun was highest in the sky. New York's day started and ended five minutes before Philadelphia's. None of this mattered too much when transport was limited to 20 or so miles a day by horse-drawn coaches and there was no real need for precise coordination. When the railways came along and journey times shortened, then it started to matter, particularly when the railways began publishing their timetables. In 1848, Britain standardised time across the mainland to Greenwich Mean Time (Irish time was set 20 minutes slower). As the pace of communication increased, there was a growing need to have a master clock to bring all the local times into line. We still tell the time locally, say 2.00 a.m. in London, knowing that it is 12.00 noon in Sydney and 9.00 p.m. in New York. But there is a standard time signal that locks all the local times together.

Under normal circumstances the mammalian circadian system probably works in the same way. All the organs are linked to a central time signal and their activities can therefore be coordinated. In exceptional circumstances this mechanism may be overridden. There had been a long-standing finding that rats whose SCN had been removed could still be entrained by the regular provision of food. In all honesty this had been viewed as something of a minor irritant. Recent work in Mike Menaker's laboratory has shown that the irritant is more than minor (Stokkan *et al.*, 2001).

The rat is a nocturnal animal and nearly all of its activity is at night, including liver function. In the wild it feeds at night. Menaker and his team fed the rats in the middle of the day. This exceptional feeding pattern did not affect nocturnal behaviour but it increased locomotor and liver function in anticipation of feeding. This suggests that food intake in itself can generate an entraining signal for the liver and perhaps other digestive organs as well. The research team suggests:

> the hierarchical method is still at least partially valid: the SCN does generate rhythmic neural and hormonal signals that influence rhythms in other brain areas, in peripheral endocrine organs such as the pineal and in behaviour. Nevertheless, circadian oscillations in the liver (and perhaps in other peripheral organs) may respond more directly to the environment.

It makes sense. Although there is a normal circadian programme that times feeding behaviour in rats, there is an advantage in being able to override this innate temporal programme when necessary.

Eighty years after Richter started to study endogenous activity in the rat, we are back to one of his central themes. He showed that an animal maintains its internal stability through its general behaviour as well as through specific and automatic physiological processes. The studies on food restriction and liver activity suggest that there is a possible role for behaviour in maintaining internal temporal structure, and perhaps the model for circadian organisation in mammals looks something like that in Figure 5.5, with feedback loops from outputs, such as feeding, acting as *Zeitgeber* signals to assist in the overall regulation of the circadian system.

Once all the pathways and loops are added it starts to look like a complicated rail network. Paul Dirac, the Nobel Prize-winning British physicist, would have thrown his hands up in horror, except that he was not given to such displays of emotion. Dirac cited mathematical beauty as the ultimate criterion for selecting the way forward in theoretical physics. He once said, 'Just because the results happen to be in agreement with observation does not prove that one's theory is correct' (Goddard, 1998), which is not something you hear very often in biology. Biologists have as much appreciation of the elegance and parsimony of nature as anyone and they like to see clear systems and patterns. But the circadian system is refusing to cooperate.

In fact it gets worse, because there are more feedbacks being built in all the time. For instance, whereas light hitting the photoreceptors in the eyes entrains the oscillator, the oscillator rhythm modulates the sensitivity of the photoreceptors. Rather than the simple linear model, the circadian system is a complex network and a perturbation at any part of the network ripples around the system.

But the achievement of locating the mammalian master pacemaker in the SCN cannot be overemphasised. One of the major goals of neuroscience research is to identify the anatomical substrates of specific behavioural control systems. This has proved frustratingly difficult in many cases. The localisation of most of circadian timing function to a bilaterally paired nucleus of some 20,000 cells in the mammalian hypothalamus, the suprachiasmatic nucleus or SCN, is one of the success stories.

6

LIGHT ON THE CLOCK

Dominated by our visual sense, photoentrainment has been regarded as yet another type of image detection, essentially no different from other visual tasks ... however, the eye has evolved specializations that enable it to extract time information from the light environment, and these specializations have nothing to do with classical image detection.

RUSSELL G. FOSTER (FOSTER, 1998)

Imagine you are a laboratory rat. You live in a comfortable, clean cage. The temperature is equably warm, there is food and water when you want and you even get stroked now and then. It is a good place to bring up a family if given the chance. You are a lucky rat because instead of being injected with all manner of things, all that happens to you is that the handling stops and you are left in a continuous dim light. Your food arrives for two hours at the same time each day. Unsurprisingly, when the food arrives you visit the feeding trough because in human parlance you are not stupid. The experimentalists would observe a 24-hour rhythm of feeding activity. But is the rat's feeding behaviour an endogenous circadian rhythm entrained by the food signal or is it a driven behaviour, merely a response to the external stimulus provided by the food? Does the rhythm in feeding come from outside the rat or inside?

If only the rat could talk. Instead, the only way to interrogate the rat is through experiments to determine whether these regular, timed visits to the feeding trough stop when the daily food 'signal' is replaced by the continuous provision of food. If the regular feeding stops, then it is likely that the food signal has acted to 'exogenously' drive a rhythm. But if the rat continues to make timed visits to the feeding trough, even when food is there all the time, and these visits free-run in the dark with an interval or period of around 24 hours, then this would demonstrate that the food has acted as a *Zeitgeber* and entrained an endogenous circadian rhythm in feeding behaviour.

Another way to check whether the external food signal is acting as a *Zeitgeber* is to expose the rat to environmental cycles that are different from 24 hours – say 23 or 25 hours – and see whether the circadian rhythm can entrain to this non-24-hour cycle with a stable phase relationship. If the food signal is a *Zeitgeber* then the rat should anticipate the arrival of food at an interval of 23 hours. And indeed it does.

However, you can only push a circadian system so far with these non-24-hour cycles. Most circadian rhythms show what is called a 'range of entrainment'. If you try to entrain a circadian rhythm to a cycle considerably outside 24 hours, say 22 or 27 hours, it may not be able to entrain with a stable phase relationship: feeding will be at irregular times and the circadian rhythm in feeding may even free-run through the abnormal cycle. The presentation of non-24-hour cycles to produce free-running rhythms is called 'forced desynchrony' and has been used to study human circadian behaviour.

Christiaan Huygens was the first person to study entrainment. The story goes that while on sea trials in an attempt to determine the longitude, Huygens, who was a poor sailor, had severe seasickness, which confined him to his cabin for two days. Two of his pendulum clocks were hanging from the same beam and, presumably with nothing better to do than feel sorry for himself, Huygens noticed how the motions of the two clocks were synchronised (Huygens, 1893):

> *it is quite worth noting that when we suspended two clocks so constructed from two hooks imbedded in the same wooden beam, the motions of each pendulum in opposite swings were so much in agreement they never receded the least bit from each other and the sound of each was always heard simultaneously. Further if this agreement was disturbed by some interference it re-established itself in a short time.*

The pendulums moved in time with each other because as the vibration of each was transmitted through the stiff beam each mutually entrained the other (Bennett *et al.*, 2002). In most living organisms it is the changing nature of light at dawn and dusk that does the entraining. This signal is 'read' by the organism and conducted through to the clock mechanism so that it is entrained or locked onto the shifting pattern of the day resulting from the earth's rotation on its inclined axis as it moves in an elliptical orbit around the sun.

In the natural environment, many 24-hour factors could act to entrain circadian rhythms, including the daily change in light, temperature, food availability, humidity or even social contact. And most of them do in one species or another. However, those factors that provide the most stable indicator of the time of day will be selected to entrain circadian clocks, and for almost all organisms the 24-hour change in the amount of light and dark provides the most robust signal of the time of day. In the absence of light, other factors may act as *Zeitgebers*, such as food or temperature, but if a temperature and light/dark cycle are presented to an organism out of phase, most will ignore the temperature cycles and entrain to the light/dark cycle. As a result, most of our understanding of entrainment has come from studies on the effects of light on the circadian clock.

Light is used by living organisms in a variety of different ways. It is the vital energy source for most of the life on earth. When reflected off objects it can be used to detect and build an image of the world for visual recognition. At the risk of stating the obvious, light can be used to signal dawn and dusk. Light can also be used to calculate the time of year. This is because the amount of light falling on the earth varies very precisely with latitude and season. If an organism can determine daylength (or night length), it can use this information to time annual events such as migration, hibernation or reproduction, a phenomenon called 'photoperiodism'.

But how is light used as a *Zeitgeber*? Consider the problem faced by a nocturnal burrow-living fieldmouse, which needs to confine its activity to the night. It has to anticipate the onset of day, otherwise it runs the risk of being caught too far from its burrow and not being able to return before dawn. If its metabolism 'fires up' too early from its daytime sleep and the mouse is ready to go before it is fully dark, it may emerge from its burrow too soon. In either case it will be a prime target for predators.

Many animals have a more-or-less standard ration of activity each day, a duration called '*alpha*'. So how can the fieldmouse use its circadian system to help align its activity period (*alpha*) to a constantly moving dawn and dusk and so avoid the problem of being a meal?

Or what about a diurnal mammal such as ourselves? The situation is basically the same except that we need to fine-tune our behaviour to the day. Our early ancestors probably slept during the night and emerged from the cave during the morning, and were adapted to function under bright light conditions. Like the fieldmouse and every other living organism, we have to adapt also to a constantly moving dawn and dusk. If the circadian system cannot track successive dawns (or dusks) then it will have little survival value in the real world where competition for resources is a life-and-death struggle.

Human beings are used to a world that they fashion for themselves. We can function as we do in the first half-hour from waking in a 'sleepy and sloppy' fashion. Fieldmice do not have this luxury, nor did our ancestors. They had to be absolutely on top of their game; there really was no room for error. In the natural world, half asleep or half awake is likely to be disastrous. In this context, the precision of entrainment is vital if the organism is to be at peak performance.

Colin Pittendrigh was one of the first to make a systematic study of entrainment. He explored the effects of short discrete pulses of light (phasic light exposure) on the free-running rhythm of animals kept in constant darkness. He showed that light exposure delivered to a free-running animal at different times in its circadian cycle has different phase-shifting effects on the observed free-running rhythm in activity. He was able to show that if light pulses are given when the animal 'thinks' it is day (subjective day), then it has little effect on the clock. However, light falling during the first half of the subjective night causes the animal to delay its activity the following day, whereas light exposure during the second half of the subjective night advances the clock (Figure 6.1). The intensity and duration of the light pulse also determine the size of the phase shift. This was first shown by Woody Hastings, working with the bioluminescent dinoflagellate alga *Gonyaulax polyedra* (Hastings and Sweeney, 1958).

Overall, light pushes and pulls the free-running rhythm towards a 24-hour cycle and at the same time ensures that the activity is appropriately aligned to dawn and dusk. The easiest way to think about the

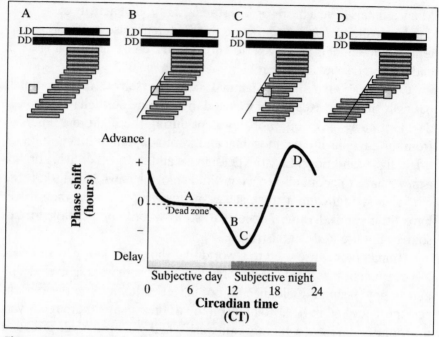

Figure 6.1. A diagram of the normal entrained 24-hour pattern of a nocturnal animal that has been kept for four days in a light/dark cycle (L:D 12:12). On day 5, the lights were switched off and the animal was kept under constant darkness; it started free-running with a period slightly shorter than 24 hours. To provide reference points to study an animal under free-running conditions, the convention is to designate activity onset in a nocturnal animal as circadian time 12 (CT 12). CT 0–12 is considered as the subjective day, and CT 12–24 is considered subjective night. But let us stress that the animal is in darkness all the time and this is simply an analytical device. If the animal gets a one-hour light pulse during its subjective circadian day (A) there is no phase-shifting effect on the free-running rhythm. This is called the 'dead zone'. At (B) the light pulse is given early in the subjective night; the effect is to make the animal start its activity slightly later the next day (a delaying phase shift). In (C) the light exposure is later into the night and there is an increased delaying effect the following day. Now look what happens at (D), when light is given during the second half of the night. *The effect is to advance the free-running rhythm (the animal starts its activity earlier the next day).* If the phase shifts (A–D) are plotted against the circadian time the result produces a phase response curve (PRC).

delaying and advancing effect of light on the circadian rhythm shown in Figure 6.1 is to consider our nocturnal fieldmouse, out in the wild, emerging from its burrow during the early evening. It is not a healthy situation for a fieldmouse to be moving about when the sun is still up. Assuming it does not get eaten, the animal will get a pulse of light that will cause it to start its activities later the next day, so it will emerge later, and thus avoid moving around in the light. Its clock will have been reset.

We will assume our fieldmouse gets away with its early emergence, but on another day it finds itself out and about from its burrow after sunrise. The effect of light will be to advance its activity earlier the next day so it will have completed its foraging before dawn arrives. In this way an organism's *alpha* is constantly being pushed back and forth so that it self-corrects around dawn and dusk. Light acts rather like the winder on the watch. If the clock is a few minutes fast or slow, turning the winder sets the clock back to the correct time. Likewise, the light signal resets the clock. It is the major input for both nocturnal and diurnal animals.

Remarkably, the phase response curve (PRC) of all organisms responds in basically the same way, with phase shifts in the first half of the night causing a delay in the activity rhythm the following day, and a converse advance in the second half of the night. The precise shape of the PRC will vary markedly between species, with some showing large delays and others showing small advances.

With diurnal animals, as with nocturnal animals, light during most of the day has no phase-shifting effect. The effect of early morning light (circadian time (CT) 0) on an animal such as ourselves is to advance the sleep/wake cycle. We will go to sleep a little earlier the next day, and rise a little earlier. At the other end of the day, light in the early evening (CT 12) will delay our sleep/wake cycle. We will go to sleep later and rise later. In this way the sleep/wake cycle can be aligned to the moving dawn and dusk over the year.

The PRC model has been very powerful in helping us think about entrainment mechanisms and it has provided a framework for understanding how the molecular components of the clock might interact to generate an entrained circadian rhythm. However, this is still a first approximation and not a full-blown description. We do not fully understand how all of these components interact to bring about entrainment. There are species-specific interactions between light intensity, the phase of light exposure, the length of the free-running period, and the size of the delay and advance portions of the PRC.

There is another complication. Mammals are not driven exclusively by their circadian rhythms. If it is still light outside, the early-rising field-mouse does not dash willy-nilly from its burrow. Light also acts directly to modify behaviour. We humans, for example, have a therapeutic dose-response to light. On a bright day with clear blue skies we feel better and that is because light itself affects our physiology and inner states (Cajochen *et al.*, 2000). Although it is obvious that the size of our pupils is directly affected by light, so too are our blood pressure, mood and attention levels. In nocturnal rodents like mice, light tends to force these animals to seek shelter and reduce activity. If it is still light, the mouse will stay down below. But if the mouse is ready to go foraging too early, its other rhythms will also be kicking in. It will want to eat and drink and so it may venture out just a little too soon or return just a little too late. The 24-hour activity cycle is not only entrained by dawn and dusk and modulated by light, as seen in the PRC, but is also driven directly by light itself. This direct effect of light on activity is called 'masking', and combines with the clock in confining activity to the period of the light/dark cycle to which the organism is adapted. The masking effect of light is seen in all organisms that have been studied so far.

If light itself, and changes in light levels, are the keys to circadian entrainment, there has to be a close functional relationship between the light-sensitive photoreceptors and circadian clocks. It was assumed that in humans and other mammals the same light-sensitive cells in the eye that were used for vision were also used to adjust the clock. But in the early 1990s, unexpected experimental results began to generate unease about this assumption.

For example, the circadian system needs relatively bright and long exposure to light to entrain compared with the visual system. The visual system of the hamster is some 200 times more sensitive to light than the circadian system, and hamsters will not entrain to a light stimulus shorter than 30 seconds. The human circadian system is more than 1,000 times less responsive to light than our visual system, and again needs a long duration signal to entrain. Further, the circadian system is sensitive to gross changes in environmental light rather than just the specific patterns of light.

Mirroring these differences, there is a very marked anatomical separation between the projections from the eye to the visual and circadian structures within the brain. The layer of light-sensing and light-processing cells in the eye is called the retina. It has been likened to a carpet, with the outer part analogous to the tuft containing the light-sensitive rod and cone photoreceptors and the inner part of the carpet, the weave, containing the cells that process the light information from the rods and cones. Rods are typically associated with vision in dim light and cones with colour vision in bright light. Information passes from the retina to the brain through the retinal ganglion cells whose projections form the optic nerve (Foster & Hankins, 2002) (Figure 6.2).

The master clock in mammals, the suprachiasmatic nuclei (SCN), receives its retinal projections from the retinohypothalamic tract (RHT), which is formed from a small number of distinct ganglion cells. These ganglion cells (around 1 per cent of the total) tend to be distributed evenly over the entire retina and send an unmapped or random projection to the SCN. Glutamate, a common neurotransmitter, carries the light information signal to individual SCN neurones. By contrast, the ganglion cells of the visual system send a highly mapped projection to the visual centres of the brain, such that a point on the retina maps precisely to a group of cells in the visual cortex. The visual system is thus able to deduce both how much light there is and where it occurs in specific regions of the environment, whereas the SCN receives only information about the general brightness of environmental light. So the mammalian eye has parallel outputs, providing both image and brightness information.

Mammals use their eyes to detect all light; if the eyes are lost, a mammal is both visually and circadian blind. It will be unable to entrain to light and will free-run for the rest of its days. However, if the eyes of birds, reptiles, amphibians and fish are removed they can still maintain an entrained circadian rhythm. They have several light-sensing 'extra-ocular' photoreceptor organs other than the eyes.

Mammals lost these extraocular photoreceptors during their unique evolutionary history. All modern mammals are derived from nocturnal, burrowing, insectivorous or omnivorous animals that were living about 100 million years ago. Primitive mammals would have spent their days in burrows and then emerged at dusk (Young, 1962). Extraocular photoreceptors, located under the skull, may not have been sufficiently sensitive to discriminate twilight changes.

Despite the fact that eye loss results in free-running rhythms in mammals, claims have been made from time to time that they have non-ocular photoreceptors. There was even a recent suggestion that humans had photoreceptors behind the knee, as bright light shone there apparently shifted human circadian rhythms. It was a charming thought but unfortunately nobody could replicate the findings. All the experimental evidence points to light entrainment of the circadian system of mammals occurring exclusively via photoreceptors within the eye. Eye loss in every mammal ever studied, including humans, results in a free-running circadian rhythm.

The sensory task of generating an image for the visual system is very different from the sensory task of collecting light for the regulation and entrainment of the clock. The previously reasonable assumption that retinal rods and cones detect light for both these forms of light sensing was a gross oversimplification. Although the entraining photoreceptors of mammals are clearly located in the eye, neither the rods nor the cones are required for this task, and there exists another photoreceptor within the eye.

The identification of the non-rod, non-cone photoreceptor resulted from a series of experiments over 10 years in Russell Foster's laboratory. The group started out by looking at mice with naturally occurring genetic disorders of the rods and cones of the eye. The experimental plan was to correlate rod and cone photoreceptor loss with a loss in sensitivity of the circadian system to light. The first mutant mice studied lacked all rods and most of the cones. This mutant strain of mice, known as the *retinal degeneration* or *rd/rd* mouse, is visually blind. But despite the massive (though not complete) loss of their rods and cones, these animals had apparently normal circadian responses to light.

These surprising and unexpected findings were met with either polite lack of interest or hostile rejection. The eye has been the subject of serious study for some 200 years, and in broad terms its function was thought to be understood. The conventional understanding was that light-sensitive rods and cones of the outer retina transduce light, and the cells of the inner retina provide the initial stages of signal processing before topographically mapped signals travel down the optic nerve to specific sites in the brain for advanced visual processing (Figure 6.2). Even though there was clear evidence of different visual and circadian pathways from the retina, including the side-shoot that linked into the

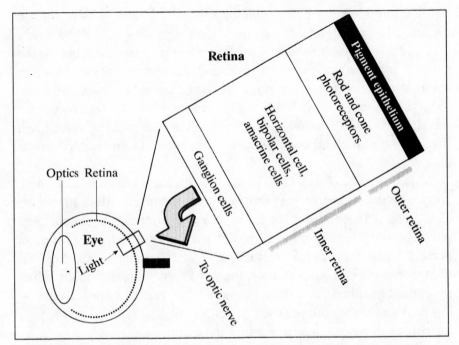

Figure 6.2. The retina of all vertebrates, including mammals, is essentially transparent and consists of three distinct cell layers. The light-sensitive rods and cones form the outer retina; the horizontal, bipolar and amacrine cells the inner retina; and the ganglion cells form a third layer. Light has to pass through the ganglion cells and cells of the inner retina before it reaches the rods and cones.

SCN, there was a strong belief among many biologists that there was a single photoreception apparatus. It seemed inconceivable that something as important as an unrecognised ocular photoreceptor could have been missed. The feeling was that there was no need to upset the neat story of vision, and its evolutionary origins, with new photoreceptors.

This conventional wisdom was challenged by a few researchers interested in how vertebrate clocks are regulated by light and whose background and training were rooted not in vision research but in fields such as circadian and reproductive physiology and animal behaviour. These researchers were not trained as visual scientists, but their *naïveté* was combined with an acute awareness that the vertebrate central nervous system is packed with 'enigmatic' photoreceptors that help adjust circadian rhythms to the local light environment. They were aware that birds, reptiles, amphibians and fish employ specialised photosensory cells in the basal brain and pineal to regulate their circadian rhythms. Foster's own undergraduate tutor was Alan Roberts, who introduced him to the study of pineal photoreception in frogs. This theme was developed in doctoral studies with Sir Brian Follett on the interplay of light and seasonal reproduction in birds. Michael Menaker, who at the time was on a sabbatical year at Bristol University, took an interest in this work and suggested that it could be developed using mice with hereditary retinal disorders.

Circadian biologists had no problems with organisms having a whole array of photoreceptors performing visual and/or circadian functions. This made it much easier for them to accept the notion of dedicated photoreceptors in the retina separate from the visual system. It seemed perfectly reasonable to ask whether rods, cones, uncharacterised retinal photoreceptors or a combination thereof might regulate the circadian rhythms of mammals, and these questions have radically altered the way in which we think about the eye as a sensory organ.

But when you are out to convince the sceptics, or trespass across the disciplinary fields erected by scientists, then the science has to be like Caesar's wife – beyond reproach. Further studies, using different mouse mutants in different laboratories around the world, were able to repeat the Foster laboratory's observations and showed that the loss of visual responses due to retinal disease did not block circadian responses to light. At the very least, these studies in mice, and other rodents such as the 'blind mole rat' (David-Gray *et al.*, 1998), showed that the processing of light by the eye for vision was very different from the way in which the eye processed information for the clock. Rodents could be visually blind but not circadian blind. However, suggestive as these early studies were, they did not conclusively demonstrate the existence of a new ocular photoreceptor. Although the rods and cones were massively reduced in these rodents, they were never completely eliminated. There was the possibility that even a small number of rods and cones might still be sufficient to maintain normal circadian responses to light.

Rather than wait for the chance discovery of a mutant mammal with absolutely no rods or cones, a new mouse strain was developed, known affectionately if unoriginally as the *rd/rd cl* mouse. This mouse had no rods or cones whatsoever. Although completely visually blind, such mice still showed the normal circadian responses to light in the laboratory. By

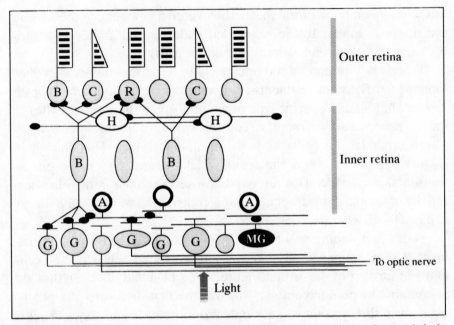

Figure 6.3. Diagram of the basic cell types of the mammalian retina and their connections. The rods (R) and cones (C) convey visual information to the ganglion cells (G) through the bipolar cells (B). The horizontal cells (H) form a plexus of neurones that allow lateral connections between the rods and cones. Amacrine cells (A) form a plexus of neurones that allow lateral connections between the bipolar and ganglion cells. The optic nerve is formed from the axons of all the ganglion cells. The melanopsin ganglion cells (MG) are intrinsically photosensitive. These few ganglion cells act as 'brightness detectors' and regulate a range of different physiological and behavioural responses to light, including the entrainment of circadian rhythms (Foster & Hankins, 2002).

blocking the light reaching the eye, the effects of light on the circadian system were abolished, so there had to be a novel photoreceptor. Because these mice lacked an outer retina, the new photoreceptor cells were probably located within the inner retina (Figure 6.3). The key questions were the nature and location of this photoreceptor, but there were no obvious candidates visible under the microscope.

Science rarely proceeds along straight lines. Chance and serendipity are the researcher's bedfellows. A clue to the mammalian photoreceptor came from work on fish. Quite by accident, Bobby Soni, who was working on his doctoral thesis, discovered a new gene in the salmon eye that was similar to the genes that code for rod and cone photo-pigments, but nevertheless had clear differences.

When photons enter a rod or cone, many of them interact with pho-topigments. The photopigments of all animals consist of a form of opsin protein that binds a specific type of vitamin A (11-*cis*-retinaldehyde) to form a photosensitive complex. The absorption of a photon of light by 11-*cis*-retinaldehyde converts it to an all-*trans* form. This change in shape of vitamin A alters the opsin, which in turn triggers the photo-transduction cascade. This ultimately causes a change in the electrical activity of the photoreceptor cell that is transmitted by a neural pathway and ends with what we know as vision.

Soni's key finding was that the new (VA-opsin) gene was not ex-pressed in the salmon's rods and cones. It was found only in certain cells in the inner part of the retina (Figure 6.3), which had not been thought to contain any photoreceptors. This was the first discovery of a photo-pigment in the eye that was separate from the rods and cones (Soni *et al.*, 1998). Although Soni had been working with fish, the discovery of the new opsin suggested a possible mechanism for non-rod, non-cone photoreception in vertebrates in general.

Rods and cones are easy to identify microscopically. Although countless microscopic investigations had revealed a huge range of cell types within the inner retina, no one had seen anything that could be considered a photoreceptor. Soni's work suggested that if the mam-malian equivalent of the salmon VA-opsin could be found and localised, there was a strong presumption that this would be the inner retinal photoreceptor.

A new opsin was soon found. Melanopsin was discovered by Ignacio Provencio in a range of mammals and localised to the ganglion cells that form the RHT and project to the SCN (Provencio *et al.*, 2000; Hattar *et al.*, 2003) (Figure 6.3). Furthermore, *in vitro*, these cells were directly sensitive to light (Berson *et al.*, 2002; Sekaran *et al.*, 2003). It looked as though these cells were the non-rod, non-cone photoreceptors. Unfortu-nately, there is so little melanopsin in the mammalian eye that it has not been possible to produce enough of the protein to do the biochemistry to prove that melanopsin is the photopigment.

An experimental conundrum soon emerged. When the melanopsin gene in a normal mouse was experimentally 'turned off', the melanopsin ganglion cells of the RHT were no longer directly light sensitive, and the circadian response to light was diminished, but critically it was far from abolished. Further, when the melanopsin photosensitive ganglion cells in the rodless-and-coneless mice are 'turned off', all circadian responses to light completely disappear (Hattar *et al.*, 2003). What has to be explained is how it is that rodless-and-coneless mice nevertheless have an apparently normal circadian entrainment to light and melanopsin-deficient normal mice have an attenuated circadian entrainment to light, but melanopsin-deficient rodless-and-coneless mice have no circadian entrainment to light. Somehow the rods and cones must play a part, but rods and cones are not necessary for apparently normal circadian responses to bright light.

Although in the bright light of artificial laboratory conditions the rods and cones are not necessary for the regulation of the circadian system, this does not mean that these photoreceptors play no role in the wild. This could be the answer to the conundrum. Under certain experimental conditions involving dim light/dark cycles that in some ways more closely resemble the light levels encountered in nature, rodless-and-coneless mice entrain but not with the same precision as normal sighted mice. In addition, electrical recordings made from the SCN of rats by Hilmar Meissl and his colleagues in Frankfurt have demonstrated that the rods and cones do send light information to the SCN (Aggelopoulos & Meissl, 2000). It could be that under bright light the contribution of the rods and cones is 'swamped' and only becomes apparent in the dimmer light of early dawn and late dusk.

Organisms have to be able to extract time-of-day information from dawn and dusk. Dawn and dusk are not single points but transient events, and during them the amount of light, the spectral composition of the light and the source of light (the position of the sun) all change in a systematic way. In theory, all these factors could be used and integrated by the circadian system to detect the phase of the solar cycle. It is possible that the changing colour of the sky at dawn and dusk, and the position of the sun with respect to the horizon, might indeed act as a cue for entrainment in some species.

Different photoreceptors, sampling slightly different aspects of the twilight scene, may allow a more accurate measure of the phase. They also allow an organism to compensate for sudden, acute changes in light environment when, say, a cloud passes over the sun, or an animal moves into the shade. Animals in the wild have to cope with both the reliably predictable daily changes and the unpredictable, moment-to-moment fluctuations.

Establishing that there are photoreceptors other than the classic rods and cones has practical applications. The studies in retinally degenerate mice encouraged studies of circadian function in blind people. Josephine Arendt and her group at the University of Surrey and Charles Czeisler's group at Harvard both identified blind individuals who had eyes but lacked conscious light perception (Czeisler *et al.*, 1995; Lockley *et al.*, 1997). Despite this, some of these individuals were able still to regulate their circadian responses to light. One practical result is that every attempt is now made to preserve an intact eye in people suffering from certain forms of eye disease so that it can perform its circadian function.

7

THE MOLECULAR CLOCK: PROTEIN 'TICK' AND RNA 'TOCK'

Although at some level everything about a simple living organism is implied in its genes, on the other hand, you really have to understand the products of the genes and how they interact, which is more complex than just knowing the sequences of the genes.

CLYDE HUTCHENSON (HUTCHENSON, 1999)

Erwin Schrödinger is best known for his metaphorical cat. The Nobel Prize-winning physicist suggested an experiment involving the homely animal as a way of thinking of some of the more outlandish ideas in quantum mechanics. The question he raised was whether the cat could be both alive and dead at the same time.

The cat in a box was just a thought experiment, the sort of thing physicists do in the bath. Schrödinger and his cat had a seminal influence on quantum physics, but the physicist was also an inspiration to the nascent science of molecular biology. His 1944 book, *What Is Life?*, considered two central issues (Schrödinger, 1992). The first is that the most characteristic feature of living systems is the ability to decrease entropy, with a consequent increase in order. Life is a continual battle to maintain a state of order in a Universe that runs towards disorder. Living systems keep climbing the entropy gradient by obtaining energy from sunlight and food. Life 1, Second Law of Thermodynamics 0 – except that the game goes into extra time and the Law gets everything in the end.

The second and, for Schrödinger, far more important issue was heredity. Aristotle had recognised that biological inheritance is about the transmission of information, but this deep insight was unrecognised for 23 centuries. It was Gregor Mendel who proposed that the information for the next generation was transmitted in discrete units that were later called genes.

Schrödinger suggested that the basis for heredity must be some kind of code, in which specific sequences of chemicals were written and interpreted. He assumed that the 'code-script', as he called it, was contained in proteins, in the form of an aperiodic crystal.

Sodium chloride or common salt is probably the most familiar periodic crystal. Once the position of one pair of sodium and chlorine atoms is known, the order and positions of all the other sodium and chlorine atoms in the lattice are defined. Regular or periodic crystals contain very little information. Schrödinger thought that the chromosome was the aperiodic crystal. Its rich information content came from the order of a small number of elements that constituted the hereditary code. He used as an example the combination of dots and dashes in Morse to show that a basic code arranged in order could contain a vast amount of information. Even at its simplest, a dot followed by a dash was not the same as a dash followed by a dot. A dot and a dash can be combined in four ways; a triplet of dots and dashes in eight ways. The combinations increase in a power law and quickly rise exponentially to huge numbers.

Schrödinger's thinking inspired a group of people to work on these problems. Among them were: Francis Crick and James Watson; Edward O. Wilson, who was to become famous for his studies on ants, ecology and sociobiology; and another young American, Seymour Benzer, who was to be a key figure in determining the mechanism of the circadian rhythm.

Benzer started out as a physicist, but in 1948 a friend passed him a copy of Schrödinger's book. Inspired by it and a chance meeting with Salvador Luria, another of the founders of molecular biology, he took a summer course in bacteriophage genetics at the Cold Spring Harbor laboratories near New York and switched disciplines (Weiner, 1999).

By the mid-1950s, biologists were groping around for the actual meaning of the gene. Benzer's aim was to link the classical gene maps that had been worked out in the first half of the century with the new molecular mechanisms suggested by Crick and Watson's discovery of DNA's double helix. In a series of experiments with what are known as rII phage mutants, he showed that the gene itself had a structure. Just as an atom could be broken into various parts, so a gene could be dissected and literally cut into pieces. The pieces were lengths of DNA. Benzer effectively showed the correspondence between the linearity of the gene and a section of DNA.

This work was vital in helping to establish the accepted view that buried within the mass of DNA of any organism is its genetic blueprint or genotype. This unique set of instructions provides the basic code for building an individual. Stretches of DNA (genes) direct the production of specific proteins, and these proteins interact with the local environment to determine how an individual will appear and function – the individual's phenotype. However, the relationship between genotype and phenotype is complex. The environment can influence the expression of the genotype and subtly alter how the DNA blueprint is read. As a result, two individuals with the same genotype may have slightly different phenotypes – 'identical twins' are not after all phenotypically identical. We can usually spot the subtle differences between identical twins with little effort.

Genes provide the blueprint for making proteins, but they do not make proteins directly. The bridge between DNA and protein synthesis is a molecule very similar to DNA called ribonucleic acid, or RNA. DNA consists of a double strand of nucleotides, whereas RNA consists of only a single strand.

You can think of DNA as a ladder. Each side of the ladder represents a chain of nucleotides, with both chains linked by the rungs. To get to the famous 1953 'double helix' of Crick and Watson, twist the ladder into a spiral. RNA is like a ladder sawn down the middle through the rungs. The nucleotide building blocks of DNA and RNA consist of three smaller molecular components, one of which is called a 'nitrogenous base' or just a 'base', for short. In DNA there are four different bases: cytosine (C), thymine (T), adenine (A) and guanine (G). In RNA, thymine (T) is replaced by uracil (U), which has the same functional properties as thymine. The two chains of nucleotides in a DNA ladder are linked together by the bases. Two bases link to form a rung. The bases can only link to each other in a very specific way: C with G, and T (or U) with A, but no other combination is possible. Thus one side of the DNA ladder is a mirror copy of the other. DNA is Schrödinger's aperiodic crystal and the list of bases in the DNA strand is the genetic code.

The building blocks of proteins are amino acids. There are 20 different natural amino acids that link together, in various combinations, into chains to produce the different types of protein and their very different activities. In the DNA strand, three bases (a codon), such as GGT, TTT or AGT, code for a particular amino acid. The sequence of codons on one side of the DNA ladder provides the code for a sequence of amino acids that will form a protein.

The DNA double helix became an iconic image in the second half of the twentieth century. Unfortunately, the representation understated the complexities of the molecule and compounded a focus on cellular heredity at the expense of cellular development and maintenance. DNA does unravel at cell division and a single strand goes to each daughter cell, where it acts as a template to form the complementary strand and so re-establishes its integrity as a double helix. But for most of the time, the cell is not dividing and neither is the DNA unravelling and splitting apart. The vast majority of the cell's time is spent in 'housekeeping tasks' involving protein synthesis and degradation. When a specific protein is to be made in a cell, only the appropriate portion of the DNA strand unzips by breaking the links between the bases (C–G and A–T). This small portion of one side of the DNA strand (half of the ladder) will then act as a template to make a complementary strand of RNA. Free nucleotides in the nucleus of the cell will line up along the unzipped DNA strand, C to G and T to U, to form an RNA strand. This single chain of RNA then peels away from the DNA strand. The result is a strand of free RNA known as messenger RNA (mRNA). The sequence of bases in the mRNA strand mirrors the sequence of bases on the DNA strand, a process known as 'transcription'. The strand of mRNA then leaves the nucleus and enters the cytoplasm of the cell. In the cytoplasm the mRNA interacts with large protein complexes called ribosomes and the information is 'translated' when amino acids line up, codon by codon, along the mRNA strand to generate a protein. This protein synthesis is also monitored for errors by other proteins (Figure 7.1).

The sequence of DNA that provides the information to make a protein can be thought of as having two parts, a coding region and a promoter region. The coding region contains the sequence of bases (codons) that defines the amino acid sequence of the protein, whereas the promoter region(s) consists of a sequence of bases that are involved in switching on the coding region. Transcription of the coding region is

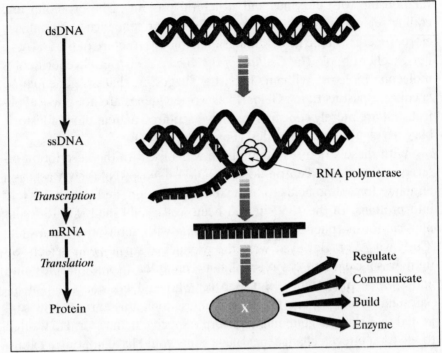

Figure 7.1. The relationship between DNA, RNA and proteins. Double-stranded DNA (dsDNA) unzips so that a single strand of DNA (ssDNA) can act as a template for messenger RNA (mRNA) transcription. Transcription begins when specific transcription factors, usually proteins, bind to the promoter region (P) of the gene. The complementary strand of RNA is synthesised by another protein complex (RNA polymerase). The strand of mRNA leaves the nucleus and enters the cytoplasm of the cell. The mRNA interacts with ribosomes, which translate the mRNA, codon by codon, into a protein (X). Proteins act as regulators, communications signals, structural elements and enzymes.

triggered when specific transcription factors, usually proteins, bind to the promoter region of the gene and stimulate transcription (Figure 7.1).

If this were a detailed textbook on molecular biology, a cautionary note would be made about the supposed simplicity of the process. The simplistic idea is of DNA as a string of beads. Each bead is a gene and each gene is responsible for one protein. This notion is far removed from any reality. DNA exists in the cell in a three-dimensional dynamic relationship with enzymes and structural proteins. The promoter and coding regions may be far apart on the linear sequence but relatively closer together in the three-dimensional folded structure of the DNA. In her excellent book *The Century of the Gene*, Evelyn Fox Keller quotes molecular biologist William Gelbart as suggesting that the gene may be 'a concept past its time'. Gelbart has written 'genes are not physical objects but are merely concepts that have acquired a great deal of historic baggage over the past decades' (Fox Keller, 2001).

With these strictures in mind, we will stick with the description offered so far and the central dogma – DNA makes RNA; RNA acts as a template for amino acids to be linked into a chain; and the chains fold into proteins. In the 1960s it had been well established that different physical forms (phenotypes) of *Drosophila* with descriptive names like 'Curly wings' or 'Bar eyes' were the product of single gene defects. So there was a clear and obvious link between at least some specific genes and specific phenotypes. But Benzer, who had become tired of his painstaking phage gene-mapping studies, took this further and suggested that a single gene might govern a specific behaviour. He wanted to see what effect a change in a single gene would have not on an organism's features – whether its legs were shorter, its eyes a different colour or, as in Mendel's peas, seeds that were smooth or wrinkled – but on what the organism did and whether its sensitivities changed. For example, *Drosophila* flies usually move instinctively towards the light, but some do not. What was the difference between the light-lovers and the dark-lovers; could it be a single genetic difference?

This was heresy at a time when it was thought that it would be far too complicated to determine the effect of a single gene on even the simplest behaviour. Using the analogy of a mechanical clock with its hundreds of parts, behaviour was thought to depend on the interaction of dozens, hundreds and perhaps even thousands of genes. Examining a single cog tells you very little about how the clock works.

A young graduate student, who was fascinated by the way in which organisms seemed to sense time, went to Benzer's laboratory to explore the possibility of genetic influence on such behaviour. Although Ronald Konopka's undergraduate thesis had been on circadian rhythms in plants, he became particularly interested in the genetic basis of Pittendrigh's observations that a clock regulated the dawn emergence of *Drosophila*. Benzer's interest was more prosaic. A notorious 'owl' himself who often worked at night ('my wife was a lark and I was owl. We'd see each other in between'), Benzer was interested in how humans synchronised their activities by time (Weiner, 1999).

The time-sensitive emergence of adult *Drosophila* flies was their behavioural 'Curly wings'. Flies hatch from their eggs as maggots, then after a period of extensive feeding the maggots develop a tough outer covering called a pupal case. Inside the pupal case the fly develops and finally emerges at dawn. A circadian clock controls this timed emergence. Populations of flies had free-running rhythms of pupal emergence under constant conditions of light and temperature. Emergence rhythms were entrained by light and were temperature compensated.

Konopka exposed flies to chemical agents that would cause random mutations in the DNA carried by the sperm of a fly. He then bred these individual flies and determined when the mutant offspring emerged from their pupal case. The sceptics said it would not work because even if Konopka found clock mutants he would be unable to find out what had gone wrong at the level of the gene. There might be thousands of mutations that could affect the clock. In their view all he was doing was making flies sick and even if there was a single mutation of the clock gene the fly might not live long enough to breed.

Konopka confounded them all and found what he was looking for at the 200th go. The first mutants were arrhythmic and eclosed at any old time of day or night. He then isolated two more mutants, one that emerged early (before dawn) and one late (after dawn). The next step was to analyse what was happening to the free-running circadian rhythms of the three types of mutant in constant darkness. Normal *Drosophila* have a free-running period a little longer than 24 hours, whereas those flies that emerged before dawn had a period around 19 hours (subsequently called S or short-period flies). Similarly, those flies that emerged after dawn had a period around 29 hours (L or long-period flies), and flies with random emergence times showed random activity (0 or arrhythmic). Crossing the mutants with wild-type flies revealed that the mutation linked to the behavioural change in the flies' clock was due to a single gene. Because mutations in this gene altered the period of the fly's circadian rhythm Konopka called it the *period* (*per*) gene. He and Benzer had tied behaviour to a gene and had identified the first clock gene to be found in any species. Nobody knew whether other genes were involved in the molecular clockwork and if so how they might interact, but they had taken the all-important first step in the 30-year journey to disassemble the *Drosophila* molecular clockwork (Konopka & Benzer, 1971).

All that genes do is code for proteins, and it is proteins that perform most of the functions of the organism. They act as regulators, enzymes, signalling molecules and structural components of the cell. If *per* was the gene, there had to be a protein. So what and where was PER and what did it do? (By convention, all *Drosophila* genes are written in lower-case italics, for example *per*, and the protein product in capitals, for example PER. *Drosophila per* is abbreviated to d*per* and the protein to dPER; mouse to *mper* and mPER and so on – the first letter of mammalian genes is usually capitalized.) In 1970 that was far easier asked than answered, because the molecular biologist's tools of gene cloning, polymerase reactions and genomic databases, were still in the future.

Initially, PER was thought to function at the junctions between cells, allowing cell–cell communication. But the first real key to understanding how PER might be involved in clock function came some years after Konopka and Benzer's work. Kathy Siwicki and Jeff Hall (another of Benzer's students) detected and localised PER within individual cells of the fly's body. They showed that although PER was present in very many different tissues of the body, a small group of cells in the fly brain (lateral neurones) and the eyes showed 24-hour rhythms of abundance in PER protein (Siwicki *et al.*, 1988).

In normal flies (*per*N), protein levels peaked early at night (around 8.00 p.m.) and then dropped to undetectable levels in the middle of the day. Furthermore, rhythms in PER peaked early in *per* short (*per*S) mutants and late in *per* long (*per*L) mutants, and no PER protein could be detected in arrhythmic *per*0 flies. So not only was PER rhythmic in the eyes and lateral neurones, but patterns of protein abundance mirrored circadian behaviour. But how was this rhythmicity being driven?

The likeliest prospect was that rhythms of *per* mRNA drove the rhythmic production of PER protein. To find out, flies were collected at different times throughout the day and night, then rapidly frozen by dropping them into liquid nitrogen. The frozen flies were then placed in a fine mesh sieve and shaken. The shaking separated the heads from the bodies and the heads fell through holes of the sieve. Having separated the heads from the bodies, *per* mRNA was then measured in populations of isolated fly heads. Sure enough, *per* mRNA showed a 24-hour cycle. The peak in mRNA occurred some four to six hours before the peak in

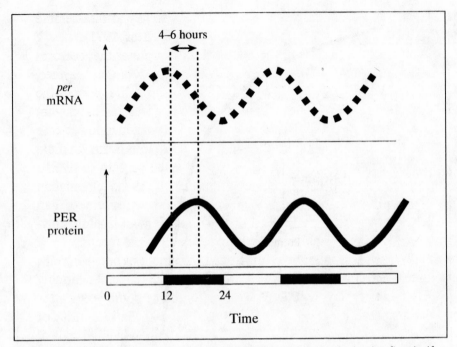

Figure 7.2. PER protein within the head of *Drosophila* peaks in the first half of the night (about 8.00 p.m.), whereas *per* mRNA peaks about four to six hours earlier.

PER protein (Figure 7.2). *per*S mutants had peaks in *per* mRNA that were advanced by approximately five hours, whereas *per*L mutants had mRNA rhythms delayed by approximately five hours compared with normal flies. Furthermore, arrhythmic *per* 0 flies showed no rhythms in *per* mRNA, and had no detectable PER protein. So mutant fly behaviour was mirrored by abnormal rhythms in both *per* mRNA and PER protein.

Paul Hardin and Jeff Hall proposed a possible mechanism for the rhythmic expression of PER in 1990, based on the idea that PER protein was involved in a negative feedback loop whereby it inhibited its own production. The idea also made sense of the crucial observation that although, like all proteins, PER is synthesised in the cytoplasm, high levels of PER protein were concentrated in the nucleus, where of course the DNA is (Hardin *et al.*, 1990).

The cycle they proposed was that *per* mRNA is transcribed from the *per* gene, allowing the production of PER protein. As PER levels build in the cytoplasm of the cell the protein enters the nucleus and inhibits the

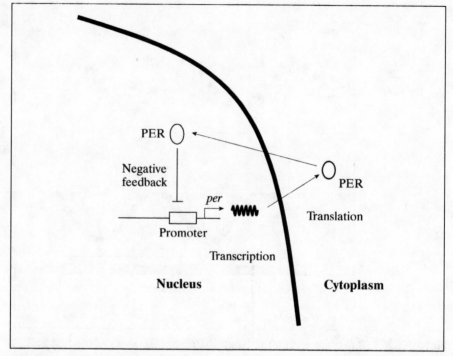

Figure 7.3. The simple 'per alone' model. The first model for the circadian clock in Drosophila was based upon a transcriptional/translational feedback loop of PER protein and per mRNA. PER protein would inhibit further transcription, and hence translation, of PER. The half-life of the protein and mRNA would determine the 24-hour dynamics of the system.

production of further per mRNA by binding with the promoter. Reduced levels of per mRNA transcription result in reduced levels of PER protein production to the point that all PER translation stops. When all PER production has stopped, and PER protein has been broken down, then this releases the per gene from inhibition, per mRNA can once more be transcribed and the whole cycle starts once again (Figure 7.3).

This is a classic negative feedback loop – but it still left unanswered the question of how this generates a stable rhythm. Negative feedback usually works to keep a system in stable equilibrium within defined limits – like the thermostat on the heating system. The nature of an oscillation is different. It describes a system that in a regular manner moves away from equilibrium before returning. To oscillate on a 24-hour or thereabouts cycle, the inhibitory effects of PER on per transcription need to be turned off, *per* mRNA production needs to be turned back on again, and there has to be a delay of many hours in the feedback process.

The dynamic tension between the positive element of the loop, promoting the transcription and translation of PER, and a negative element, inhibiting this transcription, created the rhythm. Paul Hardin wrote, 'feedback of the *per* gene product regulates its own mRNA transcription, there is a protein "tick" and an RNA "tock"' (Hardin *et al.*, 1990).

In general, proteins and mRNAs do not last very long; they are broken down and then rebuilt as and when required. The assumption was that PER and *per* mRNA would suffer a similar fate, so there was a problem with this model in arriving at delays that added up to 24 hours, but it was good for a first shot. The major problem was with dPER playing the role of a transcriptional regulator. Although it accumulates in the nucleus of the cell, dPER does not have the features expected of a protein that would bind the promoter region of a gene. In fact, dPER did not resemble any known protein. dPER does, however, possess a 'PAS domain', a 'sticky' region that allows it to bind to another protein. So if dPER cannot bind to the promoter region of DNA directly, then perhaps it binds to another protein that can bind directly to DNA. dPER needed to take a partner to the circadian dance.

Amita Sehgal and Jeff Price in Mike Young's laboratory at Rockefeller University used techniques similar to those of Konopka to identify a new *Drosophila* circadian mutant they called *timeless* (*tim*) (Sehgal *et al.*, 1995). Analysis of the *tim* gene suggested that *tim* and *per* were acting in a very similar manner. Flies that lack the *tim* gene (*tim*0) are arrhythmic. Mutations in *tim* can alter the period length of fly circadian behaviour. TIM protein accumulates in the same cell nuclei as PER protein (lateral neurones and eyes). *tim* and *per* mRNA oscillate with the same circadian phase, both peaking four to six hours before their protein levels. Critically, PER protein was never detected in cell nuclei of *tim*0 flies, and TIM was never detected in the cell nuclei of *per* 0 flies.

Furthermore, PER is very rapidly degraded in the absence of TIM protein. So when PER and TIM proteins were shown to bind to each other using their PAS domains, the pieces fell into place in a revised version of the original 'PER alone' model for the generation of a circadian clock in *Drosophila*. In this model, cycles in *per* and *tim* mRNA transcription are regulated by cycles of nuclear localisation of a PER/TIM protein complex (Figure 7.4). Although this model helped conceptualise

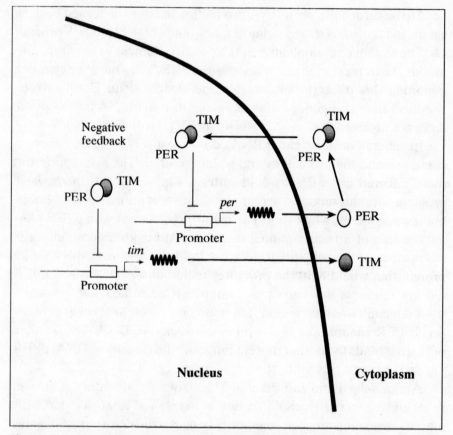

Figure 7.4. The PER/TIM model for the circadian clock of *Drosophila*. PER/TIM interaction is required for nuclear localisation of both proteins. Upon entry into the nucleus the PER/TIM complex is able to suppress *per* and *tim* gene transcription. The stability of the PER/TIM protein complex will determine the duration of transcriptional suppression of the *per* and *tim* genes. Thus the rate of entry of the PER/TIM complex into the nucleus, and how long the PER/TIM complex can persist before it is degraded in the nucleus, will determine the extent of negative feedback on *per* and *tim* transcription and hence the period of the circadian oscillation.

how a self-sustained oscillation might occur, it could not explain how the clock was entrained to the light/dark cycle. The next step was to work out the molecular basis of the phase response curve that we saw in the previous chapter.

TIM protein is stable in darkness, but it is very rapidly degraded by light at any time of the circadian cycle, irrespective of whether it is present in the cytoplasm or the nucleus. This strongly suggested a link between the oscillation of the internal circadian rhythm and the entrainment by light. For example, a light pulse during the early part of the night will normally delay the clock. Because TIM is needed for the entry of PER/TIM into the nucleus, the loss of cytoplasmic TIM during the first part of the night due to its degradation by light will result in a delayed build-up of PER/TIM and a delayed entry of the PER/TIM complex into the nucleus. This causes a delay in the suppression of *per* and *tim* mRNA. The net result will be a delay in the whole molecular feedback loop. Light during the late night has a reverse effect and advances circadian behaviour. TIM will again be degraded rapidly, but by this time TIM is in the nucleus bound in a TIM/PER complex. The breakdown of the complex releases the transcriptional inhibition of the *per* and *tim* genes earlier, resulting in an advance of the molecular cycle (Figure 7.5).

So in general terms the PER/TIM model could explain both the generation (Figure 7.4) and entrainment (Figure 7.5) of a molecular clock. But there was one nasty problem. PER alone could not bind to DNA, and neither could TIM alone; but nor could a complex of PER and TIM! So there was no way in which this complex could act as a direct transcriptional inhibitor of the *per* and *tim* genes. For a feedback loop to loop it has to close, and this one did not. The whole rationale on which the model was built was in serious trouble. However, a combination of studies in *Drosophila* and mouse clock mechanisms closed the *Drosophila* PER/TIM loop and led to the current model for the molecular mechanisms underlying the circadian clock.

In 1994, Joseph Takahashi and Larry Pinto at Chicago's Northwestern University published the results of what could have been a truly heroic study on mice. As in the studies on *Drosophila* initiated by Seymour Benzer, mice were exposed to a mutagen and the offspring of mutagenised mice were studied for circadian abnormalities. In comparison with mice, the pinhead-sized *Drosophila* breed quickly and copiously, occupy little space, and require minimal maintenance – hence their most favoured research status for the past hundred years. The problems, in terms of cost and time, of a *Drosophila*-style circadian screen on mice are magnified at least 100-fold. Takahashi and Pinto had budgeted for screening several thousand mice, but after only 25 animals

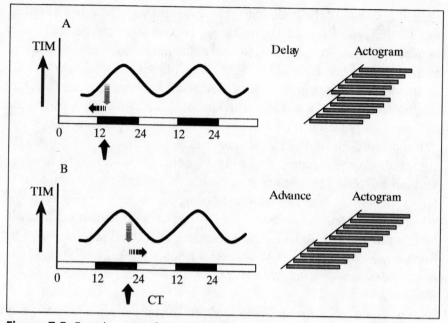

Figure 7.5. Entrainment of the PER/TIM molecular clock in *Drosophila*. (A) A light pulse during the early part of the night (grey arrow at about CT 14) will degrade TIM to levels experienced several hours earlier that day. This will result is a delayed build-up of PER/TIM, and hence a delayed entry of the PER/TIM complex into the nucleus (Figure 7.4). The actogram shows a phase delay. (B) Light during the late night (grey arrow at about CT 20) has the reverse effect, and will degrade TIM to levels that would normally occur later in the molecular cycle (Figure 7.4). TIM will be degraded rapidly and release the transcriptional inhibition of *per* and *tim* genes in advance of the normal time, resulting in an advance of the molecular cycle and an advance in behavioural rhythms as shown in the actogram.

they found a mutant mouse with a long circadian period. They called the mutant somewhat tortuously Circadian Locomotor Output Cycles Kaput, which abbreviates to *Clock*. This was the first mammalian clock gene (Vitaterna *et al.*, 1994).

At a time when the human genome is being decoded as a matter of routine, Eugene Russo was not overstating the case when he said (Russo, 1999):

By cloning the Clock *gene in mice, Northwestern University investigators helped trigger an explosion of findings on the genetics of circadian rhythms. By using forward genetics to do so, they helped introduce a new approach to doing genetics in mammals. Researchers typically perform knockout mutations on a known gene of interest in mice and then study the resulting transgenic animals (reverse genetics); the Northwestern investigators identified their gene of interest by screening for a behaviour, isolating the desired mutant, and then cloning the gene responsible.*

When mice are kept in constant darkness, *Clock* mutants with one copy of the defective gene show a lengthened circadian period, but mice with two defective copies of the gene show an even longer circadian period for a few days before becoming arrhythmic. Significantly, CLOCK protein has the necessary features to bind DNA directly and possesses a PAS domain. These features suggested that CLOCK acts as a direct transcriptonal regulator, and probably performs this function in partnership with another protein.

Once the *Clock* gene in mouse had been cloned and sequenced, it did not take long before a search of the *Drosophila* genome database showed a corresponding gene (homologue) of mouse *Clock* in *Drosophila* (*dClock*). A genome database is simply a listing in order of the bases (for humans this is three billion characters long) as they run along the DNA chain. The hard part is tying the sequencing to proteins expressed by the specific sequence lengths that we call genes. But where a coding sequence corresponds to a known gene location, it is possible to see whether there is a correspondence between the coding sequence in one species and that in another.

The *Drosophila Clock* gene protein (dCLOCK) binds (via its PAS region) to another protein that also possesses a DNA-binding region. The gene coding for this protein was called *cycle*. The discovery of *dclock* and *cycle*, and their subsequent analysis, closed the loop of the *Drosophila* clock (Figure 7.6).

CLOCK protein binds to CYCLE protein and the complex binds to a region on the promoter of the *per* and *tim* genes called an E-box element. The binding of CLOCK/CYCLE to the E-box activates the transcription of these genes and drives the production of PER and TIM proteins. After entry of the PER/TIM complex into the nucleus, it binds to CLOCK/CYCLE and inhibits their transcriptional activation of the *per* and *tim* genes. Thus PER/TIM do not inhibit their own transcription

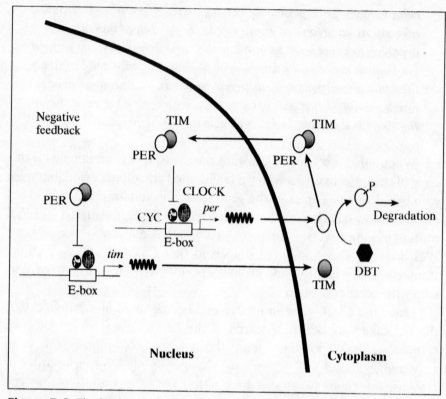

Figure 7.6. The basic molecular clockwork model in *Drosophila*. CLOCK protein binds to CYCLE (CYC) protein and the complex binds to a region on the promoter of the *per* and *tim* genes called an E-box element. The binding of CLOCK/CYCLE to the E-box activates the transcription of these genes and drives the production of *per* and *tim* mRNA, which is translated in the cytoplasm into PER and TIM proteins. After entry of the PER/TIM complex into the nucleus, they bind to CLOCK/CYCLE and inhibit their transcriptional activation of the *per* and *tim* genes. Thus PER/TIM do not inhibit their own transcription directly but do so by inhibiting the transcriptional drive of CLOCK/CYCLE on *per* and *tim*. The protein Doubletime (DBT) is a kinase that targets PER for degradation.

directly but do so by inhibiting the transcriptional drive of CLOCK/CYCLE on *per* and *tim*.

The *Drosophila* clock model explained the negative feedback loop driving the system. But usually it takes only a few minutes for transcription, translation, transportation into the nucleus and action on the nuclear targets. Yet to be of any biological use, the basic oscillation has to take about 24 hours. There had to be considerable time delays built into the feedback/expression loop and there were not enough in the model.

Part of the delay depends on another clock gene called *doubletime* (*dbt*). Its protein DBT is a kinase enzyme that works by attaching phosphate groups to dPER, which then targets the protein for destruction by the cell (Figure 7.6). dPER in its free form is phosphorylated by DBT but not when it is bound to dTIM. Until dTIM builds to sufficient levels to bind to dPER, much of the initial dPER is degraded. As the levels of dTIM rise, a stable dPER/dTIM complex is produced and it is this complex that can enter the nucleus. Inside the nucleus, it takes about 8–10 hours for the dPER/dTIM complex to break down. During this time, the transcription of dPER and dTIM is repressed and consequently there is a decreasing production of new dPER/dTIM complex and increased degradation of any free cytoplasmic dPER. After degradation of dPER/dTIM in the nucleus, the repression of the expression of dPER and dTIM is lifted, production restarts and the cycle continues.

A continual stream of new genes and proteins involved in the molecular clock are being identified in *Drosophila*. Although these genes and proteins help to explain the complex dynamics that produce a 24-hour oscillation, they certainly make a complex picture even more confusing (Hall, 2003). One of these genes in the *Drosophila* clock is *cryptochrome* (*cry*). Cryptochromes were first found in plants, where they seem to act as blue-light photopigments and have a key role in setting the plant circadian clock to the solar cycle. In *Drosophila*, CRY also sits somewhere along the entrainment pathway, acting as either a photopigment and/or part of the transduction cascade, linking a photopigment to the mechanism that causes the rapid degradation of dTIM. If CRY is a photopigment in *Drosophila*, it cannot act alone because *cry* mutant flies, which lack any functional CRY, can still entrain their circadian rhythms to a light/dark cycle. This CRY-independent light entrainment mechanism is thought to involve opsin-based photoreceptors within the clock cells of the brain and also an input from the eyes.

The current model of the *Drosophila* circadian clock in Figure 7.7 looks simple, yet it is one of the most successful genetic dissections so far of any animal behaviour. It has taken nearly 40 years of hard and often unrewarding work, but it is clear that the negative feedback loop of proteins results in a robust, self-sustaining circadian oscillation cycling at

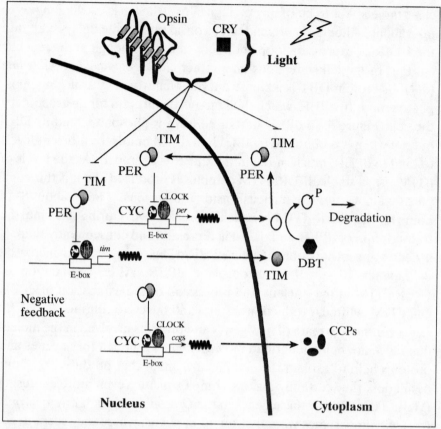

Figure 7.7. The basic molecular clockwork model in *Drosophila* with the entrainment pathway included. In this basic model of the *Drosophila* clock the PER/TIM feedback loop is aligned to the light/dark cycle as a result of light-induced degradation of TIM. Light reaches TIM via a CRY-dependent and CRY-independent pathway that involves opsin-based photopigments. The outputs from the molecular clock have been indicated as CCPs (clock-controlled proteins). Some of these clock-controlled genes (*ccgs*) are thought to be regulated via an E-box and a CYC/CLOCK-driven transcription.

around 24 hours that is entrained to the solar cycle. Although the detailed kinetics and the biochemical action of each of the *Drosophila* clock components are not yet finalised, the fundamentals of the loops and the relationships are generally agreed. We understand in some detail how the dynamic interactions of the various factors enable a cell effectively to tell time. It is a remarkable feat (Hall, 2003).

Whereas the genomes of several microbes have been determined, the *Drosophila* genome is one of only a handful of larger organisms to have been published so far, including yeast, a nematode worm, a tiny weed, *Arabidopsis*, and the mouse. The draft of the three billion codons in human DNA has been issued with the suggestion that this represents about 30,000 genes. The most recent genomic publication is the puffer fish (*Fugu*). The puffer fish code has about one-eighth the number of codons in humans but this represents about the same number of genes. Much of the human genome is regarded as 'junk' DNA whose function is uncertain, the relic of nearly four billion years of evolution.

The genomic database is messy. The long slog now under way is to work out where the punctuation marks are, at the same time as sorting real junk from the code. When it is done, it should be possible to follow the path of known genes between species and back and forth through evolutionary time.

The mouse genome is fairly well characterised and, using the *Drosophila* model of the clock genes as a guide, a reasonably comprehensive picture of the mouse clock mechanism has also been developed (Figure 7.8).

Three versions of *Drosophila per* have been identified in mice, called *mPer1*, *mPer2* and *mPer3*. All three of these genes are rhythmically transcribed with a period of approximately 24 hours under constant conditions. Although there are three versions of *per* in mammals, there may not be a true *tim* homologue in the mouse. In contrast to *dper* in *Drosophila*, the mouse *Per*s are high during the day. Also, unlike *Drosophila* in which *dper* is insensitive to light and dTIM is degraded by light, transcription of *mPer1* and *mPer2* is induced rapidly by light. The mouse homologue of *Drosophila cycle* was found by searching the database using only the PAS sequence and identified a gene already named *Mop3*. This gene is now known as *Bmal1* and has been shown to have a robust circadian rhythm of expression in the SCN.

The part that TIM plays in mice is still being determined. Recently Martha Gillette has demonstrated that a particular form of TIM seems to be a vital component of the molecular clock and might be involved in the entrainment pathway along with PER1 and PER2 (Barnes, 2003). Another significant difference is the role of CRY. In the mouse, CRY is not involved in the light input pathway but it is part of the feedback loop itself. Circadian rhythm generation in mammals is based on

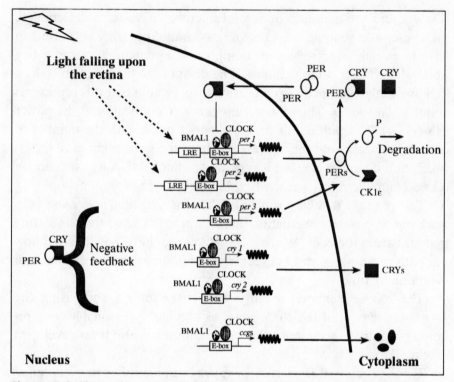

Figure 7.8. The molecular clock of mammals. Three *per* genes and two *cry* genes are thought to form the primary components of the autoregulatory negative feedback loop that constitutes the molecular clock of mammals. It is possible that PER2 acts as a positive transcriptional regulator of BMAL1, which combines with CLOCK to transcriptionally drive the *pers*, *crys* and clock-controlled genes (*ccgs*) via an E-box element (open box). Casein kinase 1ε (CK1ε) adds phosphate groups to PER proteins, which target these proteins for degradation. *per1* and *per2* genes are upregulated by light, and have light-regulatory elements in their promoter (E-box). See the text for details.

the transcriptional regulation of two sets of genes (*Per*s and *Cry*s) by an autoregulatory feedback loop. Transcription of these genes is driven by CLOCK and BMAL1, acting on E-box sequences in their promoters. In time, a build-up of the *Per* and *Cry* mRNA is followed by an increase in cytoplasmic concentrations of PER and CRY proteins. These proteins are likely to interact in some as yet unknown combination and enter the nucleus, where they inhibit CLOCK:BMAL1-induced transcription of their own genes. As with *Drosophila*, mouse DBT (casein kinase 1ε,

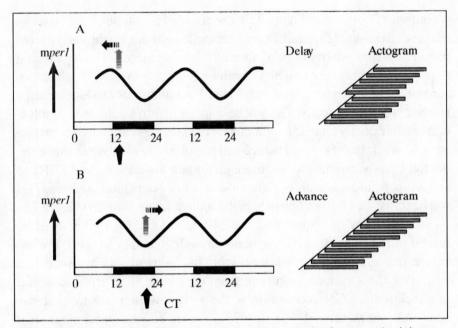

Figure 7.9. Entrainment of the mouse molecular clock of mammals. (A) A light pulse during the early part of the night (grey arrow at about CT 14) will upregulate *mPer1* to levels experienced several hours earlier that day. The actogram shows a phase delay. (B) Light during the late night (grey arrow about CT 20) has the reverse effect, and will upregulate *mPer1* to levels that would normally occur later in the molecular cycle, resulting in an advance of the molecular cycle and an advance in behavioural rhythms as shown in the actogram.

CK1ε) is a kinase that adds phosphate groups to PER protein, which targets these proteins for degradation (Figure 7.8).

The molecular clockwork of the mouse is at least as complex as that of *Drosophila*. New components that fine-tune the basic mechanism are being discovered on a regular basis. The newly discovered *Rev-Erbα* and *Dec1* and *Dec2* are thought to fine-tune the transcription of *Bmal1* to inhibit its expression. However, the fundamental mechanism of a negative feedback loop still holds. Whether the molecular mechanism of the SCN clock in the mammal is precisely the same clockwork as in the peripheral oscillators of the liver, heart and lung is as yet unknown.

The entrainment of the mammalian molecular feedback loop to the solar light/dark cycle is exclusively through receptors within the retina and their action on *mPer1* and *mPer2* (and perhaps *mTim*). Compare, as we did with *Drosophila* (Figure 7.5), the effect of a light pulse at two times of night in a mouse (Figure 7.9), and start with light during the early part of the night, a time when light will normally delay the clock. During the first part of the night *mPer1* and *mPer2* will be relatively inactive. Light exposure will activate these genes and drive PER1 and PER2 to levels seen a few hours earlier that day. This will be translated into a delay in the molecular oscillation. By contrast, light during the late night advances circadian behaviour. Both *mPer1* and *mPer2* will be relatively inactive at this time, but light exposure will activate these genes and drive PER1 and PER2 to levels that would normally be encountered a few hours later: at dawn. This will be translated into an advance in the molecular oscillation (Figure 7.9)

To be of any biological use, the 24-hour molecular cycle has to be turned into a signal that can regulate physiology and behaviour. For example, the SCN shows a 24-hour rhythm in electrical activity that drives output rhythms such as melatonin from the pineal. The problem is that this melatonin–SCN connection is the only output pathway that has been fully characterised (Klein, 1993). Yet we know that all manner of cellular and physiological activity is under clock control. Within the SCN there is a group of what are called 'clock-controlled genes' or CCGs (Figures 7.7 and 7.8), which are driven by the molecular oscillation. This rhythmic transcription seems to involve the same basic elements that drive the molecular feedback loop, with a CLOCK/BMAL1 transcriptional drive via an E-box element, and transcriptional inhibition involving PER and CRY (Figure 7.8).

One of these CCGs is the gene for the neuropeptide arginine vasopressin, normally shortened to vasopressin or AVP. In the SCN, AVP has a strong circadian rhythm in both its mRNA and protein abundance. AVP directly applied to an isolated SCN will increase the electrical activity of many SCN neurones. In addition, SCN neurones produce AVP and release it in a rhythmic manner to alter the activity of cells outside the SCN. So AVP acts as both a local and a medium-distance output signal from the molecular loop (Buijs & Kalsbeek, 2001). But AVP cannot be the only important output signal released by the SCN, because rats carrying a mutation that prevents them making AVP still show apparently normal circadian rhythmicity. Significantly, AVP-deficient SCN transplants are able to restore rhythmicity in SCN-lesioned hosts. There must be other signals that have yet to be found.

Several of the genes involved in the circadian clock are very similar in both the mouse and *Drosophila*, suggesting that there was a circadian clock in the common ancestor of insects and mammals some 700 million years ago. If we look back beyond that, before the animal branch of the evolutionary tree, and examine the molecular basis of the clock in plants, fungi or even photosynthetic bacteria, what degree of conservation will we find in our even more ancient cousins?

The answer is: not a lot. Detailed studies on the fungus *Neurospora* have identified several genes (*frequency, white collar-1* and *white collar-2*) that have been shown to be critical for the generation of circadian rhythms. However, these genes bear no real resemblance to any of the known clock genes in animals. They are not homologous. This is also true for the *clock* genes isolated from plants (for example *cca1* or *ccr2*) and photosynthetic bacteria (for example *kaiA, kaiB, kaiC*), leading to the suggestion that the biological clock may have evolved multiple times during the course of evolution.

Although different sets of genes seem to generate the clock in animals, plants, fungi and bacteria, they use the same fundamental mechanism – an 'auto-regulative negative feedback loop' involving several genes. Compare bicycles, cars, trains and tanks. They use different methods but all of them move from A to B by virtue of a rotating wheel. Similarly, the clock generates a sustainable rhythm, but different organisms use different sets of genes to the same effect. These genes give rise to a message (mRNA) and a protein that may cycle in a circadian manner and are considered so-called 'state variables' of the molecular oscillation. The protein acts either directly or indirectly as a transcription factor, downregulating or inhibiting its own gene expression.

Although the forest that is the molecular clock is clearly visible, the trees still have to be seen. In all the models we have for the generation of this molecular oscillation, important features of this feedback loop still have to be worked out in detail. What parts of the feedback loop contribute to the delays that give rise to 24-hour rhythms? Can proteins be altered to change their activity to modify rhythmicity? What are the mechanisms allowing entry of proteins into the nucleus? How are the different components degraded? How many components are really involved in generating a clock? There is many a doctoral thesis still to be written.

By 2004, the basic model for the molecular clock had been worked out. The genes produced proteins and the cycling proteins produced a rhythm. Or so it seemed. Not everyone is convinced by this somewhat mechanical explanation that lends itself to complicated but relatively self-contained charts of genes and their products in interacting loops.

Till Roenneberg of the University of Munich argues that Colin Pittendrigh was right to talk about a circadian system rather than a clock. He says (Roenneberg & Merrow, 2001):

> the circadian system has a lot of qualities; one of them is rhythmicity, another one is circadian periodicity, next one is self-sustainment, another one is amplitude robust enough to drive output rhythms, yet another is temperature compensation and the sixth is entrainability. These qualities might come from very different elements. One element might generate rhythmicity which is not in the circadian range; it could be metabolic and not temperature-compensated which means it will be sloppy and not very robust and other elements which themselves do not generate rhythmicity might add qualities to the circadian system. We should not believe that because we look at the circadian system as one set of qualities that they all belong to one set of molecules when we go into the details of the system. They could all contribute to different molecular functions within the clock as well as to different functions within the cell.

Roenneberg's essential point is that we tend to find what we are looking for. We impose an order on our observations that fits the data into a tidy theoretical perspective. Look for clock genes and you find clock genes. When an organism is kept in constant darkness its rhythm 'free-runs'. But in nature, organisms are almost never in constant darkness and Roenneberg's pertinent question is: how can it be, in a world of dawn and dusk – in other words of only entrained conditions – that the system evolved to oscillate under constant conditions?

His answer, which is explored further in the chapter on the evolution of the clock (Chapter 10), is that while the one-way input pathway, rhythm-generating loop and unidirectional output pathways have formed an excellent scheme for finding critical components within the circadian pathway, at both the anatomical and the molecular levels, this scheme is of limited use. What we call a clock gene may have an important function within the system, but it could be involved in other systems as well. Without a complete picture of all the components and their interactions, it is impossible to tell what is part of an oscillator generating rhythmicity, what is part of an input, and what is part of an output. In a phrase, it ain't that simple!

8

A FEW SPECIES AND
MANY CLOCKS

We have many miles to go in cataloguing all of the earth's organisms
DAVID GIANNASI, UNIVERSITY OF GEORGIA (PALEVITZ, 2002)

Nobody knows how many species there are on this planet. Estimates vary from two million to 100 million, with the best estimate somewhere near 10 million. But it could be more. In one study a few years ago of just 19 trees in Panama, 80 per cent of the 1,200 beetle species discovered were previously unknown to science (UNEP, 1995).

Aristotle made the first systematic classification of living things by dividing them first into plants and animals. He then grouped animals into Land Dwellers, Water Dwellers and Air Dwellers and plants into three categories based on differences in their stems.

Aristotle's classification lasted for some 1,600 years. Although Linnaeus is credited with introducing the simple hierarchical system for classifying and naming organisms in the mid-1700s, he was preceded a hundred years earlier by John Ray.

Ray was a genuinely remarkable individual and a salutary reminder that among the early scientific investigators were not only the curious but also the brave. Born in a smithy, he graduated from Trinity College, Cambridge, and became a minor Fellow in 1649. While a lecturer, he preached regularly at chapel in Cambridge, but his ordination was delayed until 1660 by the disruption of the Civil War. During the war, many ministers had signed a Covenant – a manifesto for reform of the Church. In 1662 the new king, Charles II, required every minister to swear an oath condemning the Covenant. Although Ray had not himself signed the Covenant, he would not condemn it, and accordingly lost his university post, his house and his botanic garden.

He was left with no income and no place to live, but past students and their wealthy friends gave him various jobs tutoring their children. In his spare time he worked unpaid in the natural sciences and produced the first botany textbook and a systematic classification of plants (Baldwin, 1986).

The familiar classification, pioneered by Ray, of Kingdom, phylum, class, order, family, genus, species ('King Philip Came Over For Grandma's Soup') for the huge number of species is the key to biology. Darwin's towering achievement was to discover a unifying theme that linked species, genus, family and so on into an evolutionary whole. But he had to spend five years on the *Beagle*, in the company of a depressive captain, gathering a great many observations and then another 20 years turning his mass of observations and thoughts into *The Origin of Species*. It was a wonderful feat of inductive reasoning, of arguing from the particular to the general.

His simple concepts provided the driving theoretical perspective for biological understanding. As the geneticist Theodosius Dobzhansky succinctly put it, 'Nothing in biology makes sense except in the light of evolution'. Circadian research is no exception. Understanding the organisation of the circadian system in different species is the key to understanding circadian systems in general, from the molecular to the ecological.

In the 1960s, birds were the mainstay of circadian research. The common house sparrow was a favourite laboratory animal in the USA because, as it was a designated pest, sparrows were free and what is more they were tough.

Mike Menaker, working at the Universities of Texas and Oregon, used sparrows for a classic set of investigations spanning the 1970s on the regulation and generation of the circadian system of birds. In the first experiments, he showed that house sparrows with no eyes could still entrain their circadian rhythms of perch-hopping activity to a light/dark cycle, but they needed a brighter light than sighted sparrows. So, unlike mammals, the eyes of birds do not provide the only light signal to the circadian system. His informed guess was that these photoreceptors reside within the brain.

Menaker then took the eyeless sparrows and exposed them to a dimmer light/dark cycle. The birds did not entrain and, after several weeks of free-running perch-hopping activity, the feathers on the top of the head were removed, allowing more light to penetrate into the brain. The birds were then able to entrain to the dim light/dark cycle! There was a threshold light level, and light brighter than the threshold could be detected and the circadian rhythm entrained. After a further few weeks, the feathers grew back and the birds showed free-running rhythms once more. However, after another round of plucking, entrainment returned. His guess was correct: the photoreceptors regulating circadian entrainment must reside somewhere within the brain (Menaker, 1972).

The idea that photoreceptors are covered by bone and tissue sounds odd, but Buddhist monks determine the time to rise when it is light enough to see the veins in the hand held up to the sun. Large amounts of light can penetrate deep into the body and easily pass through the lightweight, translucent skull and brain of a bird. We prove that as youngsters when we crawl under the bedcovers and press our palms over the beam of a torch.

The next step was to remove the bird's pineal organ. The pineal develops from a similar region of the brain to the eyes, and the pineal and eyes have many features in common in non-mammals, including photoreception. In non-mammals the pineal is situated near the surface of the brain, and light reaches the pineal by passing through the skin and skull. In some fish, however, there is a semi-transparent area or 'window' that allows even more light to reach the pineal.

Menaker was building on previous studies by Eberhard Dodt, who had shown that the pineal systems of fish, frogs and lizards respond to light and send neural messages to the rest of the brain, and Andreas Oksche, who had noted in these animals that pineal photoreceptors were strikingly similar to the cone photoreceptors of the retina.

When Menaker removed the pineal from blind house sparrows that had been kept under a light/dark cycle, surprisingly, the pinealectomised birds remained entrained! Birds with neither eyes nor a pineal could still entrain to a light/dark cycle. There had to be a non-eye, non-pineal, brain photoreceptor as well, and subsequently deep-brain photoreceptors were fully characterised in studies on the regulation of the seasonal reproductive cycles of birds, as we will see in Chapter 10.

Menaker then had a stroke of luck. The light/dark cycle in the light-tight boxes was accidentally turned off and the pinealectomised house sparrows were plunged into constant darkness. In a few days the rhythmic activity that was being entrained by the light/dark cycle degenerated and broke down completely. Birds with no pineal had become arrhythmic under constant conditions. Menaker reasoned that although the light/dark cycle could sustain rhythmic behaviour in a pinealectomised bird, the remaining parts of the circadian system were not sufficiently robust, or self-sustained, to maintain rhythmic behaviour under constant darkness for more than a few cycles. The pineal itself might not be the only oscillator of the sparrow's circadian system, but it was almost certainly the major one.

The pineal makes direct connections with the brain down the pineal stalk. However, when all outputs from the pineal were cut, behavioural rhythmicity in constant conditions was not abolished, suggesting that there was a non-neural and almost certainly hormonal signal from the pineal to the brain. The pineal was known to produce and rhythmically release the hormone melatonin. Menaker's initial thoughts were that the sparrow circadian system consisted of multiple oscillators that were synchronised by a dominant oscillator in the pineal through the rhythmic release of melatonin.

He and his colleagues tried to transplant pineals back into the pineal chamber of birds that had been pinealectomised, but with very limited success. They had better luck when they implanted the pineal tissue into the eye. It may seem an odd choice, but the anterior chamber of the eye is used for transplantation studies because it is well supplied with blood and is immunologically protected. A few days after transplantation, the arrhythmic recipient bird showed robust rhythmic activity. Not only did the transplanted pineal confer rhythmicity but, because neural connections take longer than a few days to become established, the rapid restoration of activity suggested that the output from the pineal driving rhythmic behaviour was hormonal (Zimmerman & Menaker, 1979).

In the final experiments two groups of birds were entrained to light/dark cycles that were 180° out of phase (that is, 12 hours apart). The pineals were removed from these birds and transplanted intraocularly into arrhythmic hosts. The phase of the restored behavioural rhythm was dictated by the phase of the donor bird. So in 1979 Menaker had shown that the transplanted pineal not only conferred rhythmicity to the host bird but also determined the phase of the restored behavioural rhythm. The pineal had to contain a clock because the transplanted pineal carried with it a critical component of any clock, its phase (Figure 8.1). This was the first time that a circadian clock had been clearly localised to a part of the brain in any vertebrate (Zimmerman & Menaker, 1979).

But if the pineal was the master clock in the sparrow, that still did not explain how it was that a pinealectomised bird remained rhythmic when it was kept in a light/dark cycle. Work on the mammalian SCN had strongly suggested that it contained a circadian clock and Menaker reasoned that there might be an oscillator in the house sparrow SCN. Sure enough, when the house sparrow SCN was destroyed, behavioural rhythms disappeared instantly, even when the pineal was present.

As we described above, when the pineal was removed, the house sparrow could not maintain its circadian rhythm in constant conditions for more than a few cycles. The bird's SCN is what is known as a damped oscillator and does not produce a self-sustaining rhythm, unlike in mammals. The sparrow circadian system is critically dependent upon both the pineal and the SCN to maintain a self-sustaining rhythm.

If the pineal of a sparrow is removed and placed in culture medium not only can it be kept alive for many days but under constant darkness conditions it will continue to produce a rhythmic output of melatonin that can be entrained by a light/dark cycle. Even when a pineal is separated out into individual cells, each one is still capable of producing a rhythmic output of melatonin. The direct link between melatonin and behaviour was established by infusing a pattern of melatonin into the blood of free-running birds to mimic a light/dark cycle and demonstrating that the rhythms became entrained.

The pineal, melatonin, deep-brain photoreceptors and the SCN all play key roles in the organisation of the circadian system of the sparrow. But if sighted sparrows are held under a light/dark cycle at an intensity that is just bright enough to allow entrainment, and then the eyes are either removed or shielded from light, then at this threshold light level the birds will free-run. This suggests that the eyes add to the overall sensitivity of the circadian system to light.

The isolated retina of the sparrow produces melatonin that is released in a circadian pattern, so the sparrow's eyes must contain an independent clock! This rhythmic melatonin can certainly reach the

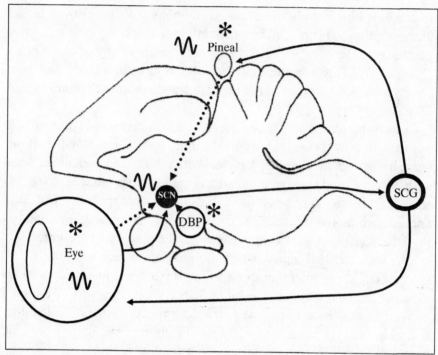

Figure 8.1. This diagram of the house sparrow brain sectioned down the mid-line, showing the cut surface of the left-hand half of the brain, illustrates the key structures and connections that seem to be involved in the generation and regulation of circadian activity. The pineal or 'master pacemaker' contains photoreceptors that entrain a circadian rhythm of melatonin release (dotted line). The pineal also receives a circadian input from the SCN by way of the superior cervical ganglia (SCG) and the sympathetic nervous system. The eyes provide light information to the SCN via a direct neural connection (the retinohypothalamic tract). The eyes also provide the SCN with a rhythmic, clock-driven, melatonin signal (dotted line). The SCN might also in turn regulate the eye via the sympathetic nervous system and the SCG. Light information to the SCN is also provided by the deep-brain photoreceptors (DBP). Stars indicate photoreceptors; sinusoidal waves indicate circadian oscillators.

bloodstream, and as melatonin has been shown to influence circadian behaviour this may also contribute to the overall circadian organisation of the sparrow. Just like mammals, birds have a retinohypothalamic tract, which links the retina to the SCN. The eyes might also receive information from the SCN as a result of inputs from the sympathetic nervous system. So the SCN might be part of a loop whereby it rhythmically regulates the eyes, which in turn send a rhythmic signal back to the SCN.

Based largely on findings in the sparrow, a model for circadian organisation in birds emerged and has been described by Vincent Cassone and Mike Menaker as the 'Avian Neuroendocrine Loop' (Figure 8.1).

Clearly, the circadian system of the sparrow is complex. There are multiple pacemakers and multiple photoreceptors. But can we infer that the organisation of the circadian system of the sparrow is typical of all birds? Can we generalise from the sparrow and talk about not only birds but other animals such as reptiles?

It is not that easy. Remove the pineal from a Japanese quail or a chicken and there is little or no discernible effect on circadian behaviour. Pinealectomy of starlings or pigeons will cause variable levels of disruption to behavioural rhythms, whereas loss of the pineal in house finches, and Java sparrows, will cause arrhythmicity. Removal of the eyes causes arrhythmicity in the Japanese quail, and a marked disruption of behaviour in pigeons. Combined eye loss and pinealectomy in the pigeon results in arrhythmicity, but some rhythmicity is maintained in starlings that have lost both their eyes and their pineal!

How can there be such startling differences between birds? After all, the whole point of classification is that fish go with fish, birds with birds, and so on. There is enormous variation, but the basic design ought to be somewhat similar. Ebo Gwinner and Roland Brandstatter have made the point that while there is a common design, birds are ecologically very diverse with very complex life strategies, and perhaps this is somehow reflected in the organisation of their circadian systems. They argue that the circadian system of migratory birds has to provide information about the time of year and has to be rapidly adjusted so as to prevent 'jet-lag' during migration. Non-migratory temperate-zone birds have to cope with marked changes in daylength and *Zeitgeber* strengths over the year. Perhaps a circadian system composed of multiple, but interacting, oscillators and photoreceptors tuned to different aspects of the light environment allows birds to adjust more efficiently to a highly dynamic time environment? But this would only partly explain why pacemakers in the pineal or eyes play a dominant role in some species but not in others (Brandstatter, 2002). We simply do not know.

The only common feature of circadian organisation in birds is that destruction of the SCN will always cause instant arrhythmicity. This suggests that at the very least the pathway from the SCN that drives rhythmic behaviour has been conserved during evolution.

The circadian organisation of fish, amphibians and reptiles shares many of the basic characteristics seen in birds. They possess circadian oscillators in the pineal and eye that regulate a rhythmic output of melatonin. They possess multiple photoreceptors that allow circadian entrainment, and they possess an SCN-like structure that houses a circadian oscillator and receives inputs from multiple photoreceptors. Like birds, the importance of these structures varies markedly in even closely related species, but a real understanding of how these structures interact to drive rhythmicity remains poorly understood. A major reason for this is that it has been technically difficult to record robust circadian rhythms from such animals. They do not like to run in a running wheel or jump up and down on a perch.

Steve Kay, who is now at the Scripps Institute, solved a similar experimental problem in plants. It was hard to find a rhythm that could be easily monitored. Leaf movement is all well and good but is difficult to automate in large numbers of individual plants. The leaves of an individual plant can be 'wired up' but to do this in large numbers is impractical. Steve Kay and Andrew Millar genetically engineered *Arabidopsis* plants so that firefly luciferase served as a marker for the clock. The tiny weed is a cousin to the mustard that goes with cress, and when Kay wants to tell the time he sprays some of the specially prepared tiny plants with a fine mist of luciferin, the small organic molecule that gives the firefly its fire. The plants begin to glimmer a few hours before dawn, growing steadily brighter through the morning and weaker as the day wears on. This neat way of giving the plant's clocks a convenient pair of hands was a lot easier than monitoring leaf movements for days on end and has turned *Arabidopsis* into the botanical world's mouse for the purposes of genetic investigation of the molecular clockwork.

The same approach has been used in *Drosophila*. When the luciferase gene is linked to part of the *per* gene in fruit-flies, there is a glow in the heads and eyes of the flies when the gene is turned on by the clock. There is also a glow in the flies' bodies, and even out near the tips of their wings. The firefly's gleam showed that fruit-flies and most probably all animals have molecular clocks in virtually every cell of their bodies.

And, stranger still, these clocks keep on ticking even when disconnected from the fly's brain.

Circadian oscillators in vertebrates were thought to reside exclusively within discrete regions of the central nervous system, such as the SCN of mammals or the pineal of birds. However, a variety of non-neural tissues have now been shown to express circadian rhythms when isolated from the SCN. As we discussed in Chapter 5, mammalian fibroblasts or whole organs in culture show 24-hour cycles of clock gene expression for several cycles before rhythms damp out and disappear. In these mammalian studies, light was incapable of entraining the patterns of clock gene expression, supporting the notion that mammalian photoentrainment is mediated exclusively by the eyes. But, once again, we cannot make generalisations about the vertebrates based upon studies in mammals.

Recent studies in zebrafish by David Whitmore, Nicholas Foulkes and Paolo Sassone-Corsi show a more advanced pattern of circadian decentralisation than is seen in mammals. Individual non-neural organs and tissues from zebrafish show robust 24-hour patterns of clock gene expression and, unlike in mammalian tissues, these rhythms will continue for many cycles in culture without damping out. Perhaps more surprising, however, is that these rhythms in clock gene expression seem to be entrained directly by light. As a result, the peripheral organs, tissues and cells of zebrafish (like *Drosophila*) must contain a photopigment and signal transduction cascade that is capable of mediating these effects of light on the molecular machinery of the clock (Whitmore *et al.*, 2000). At the moment these non-neural tissue photoreceptors remain completely unidentified in all animals. However, many researchers predict that tissue photoreception will be found in all small, relatively transparent animals, where light can penetrate deep into the body.

Zebrafish may well become the key organism for studying circadian rhythms, rivalling *Drosophila*, mice and *Arabidopsis*. With the sequencing of the zebrafish genome, it is clear that many genes in mammals, including clock genes, have homologues in zebrafish so it is possible to see whether they act as clock genes in the fish. Further, Gregory Cahill at the University of Houston has established an on-line computerised image analysis system to measure the circadian rhythm of the swimming activity of larval zebrafish. These baby fish show rhythmic behaviour for more than a week, and Cahill can monitor up to 150 fish simultaneously at very little cost. He has found that pinealectomy or removal of the eyes does not alter the entrainment or the period of free-running rhythms. Both the pineal and eyes produce rhythmic melatonin, but they do not seem to be vital for the regulation of circadian swimming behaviour. Cahill suggests that other oscillators, probably the fish SCN, are responsible for the regulation of behavioural rhythms (Cahill, 2002).

Gradually, observations on circadian organisation are being made and gathered from bacteria, single-celled organisms, plants and a wide variety of invertebrates and vertebrates. Andrew Millar at Warwick University has pointed out that the formal properties of circadian rhythms – persistence in constant conditions with an approximately 24-hour period, entrainment, and temperature compensation – are extraordinarily similar across the kingdoms of the living world: 'This initially seems paradoxical, because the molecular, metabolic and behavioural processes that they control are tremendously diverse. Similarly, entrainment to the day/night cycle is critical to every true clock, but we know that the photoreceptors involved vary widely among organisms.'

Millar considers that we have to move on from the simple input–oscillator–clock model that has dominated thinking on circadian rhythms for the past 40 or so years. In his view (Millar, 1998),

Unfortunately experimental systems do not provide a circadian clock in such splendid isolation but rather in the context of a network of interacting signal transduction pathways. The signalling network has profound effects on the oscillator, such as modifications of the free running period, suggesting that it will be extremely difficult to understand the oscillator in vivo *without some knowledge of the network's organisation.*

Towards the end of his life, Pittendrigh talked about a circadian system rather than a circadian clock. He was moving towards the network idea with signals being passed back and forth between the constituents of the network in an endless sequence of loop and counter-loop. He was also clear that if we are to understand this system it has to be examined by the one all-powerful explanatory tool in biology – evolution. It may be the only way to make sense of circadian organisation but it is going to be a long trek to a general theory (Pittendrigh, 1993).

9

THE CHANGING SEASONS

They're patterned from ancient times. When they decide they want to move, they just turn and go. We don't have to persuade them.
JOHAN EIRAP, A LAPP HERDER EXPLAINING HOW HIS REINDEER ALMOST
ALWAYS MIGRATE TOWARDS THEIR WINTER HOME ON THE SAME DAY
EACH YEAR – 29 JULY (HODGE, 2001)

Each year, as the winter frosts move in around the northeastern USA and eastern Canada, monarch butterflies set out on a 3,500-kilometre journey to the mountains of central Mexico. Like Johan Eirap's reindeer, these insects know when to leave and then navigate with remarkable precision for thousands of kilometres to a patch of land a few metres across (Froy *et al.*, 2003).

On land, sea and air, animals are continually on the move, searching for the best conditions in which to mate and for their progeny to survive. Whales travel halfway around the world's oceans to mate, and salmon cross huge areas of water to return to the exact spot where they were born; in the air, birds cross whole continents in their search for food and a safe place to breed.

Biologists are fascinated by how they do it. Our ancestors were just as fascinated. In the Grotte de Font-de-Gaume, a cave in the Dordogne, early Cro-Magnon men carved and painted images of herds of bison, horse, mammoth and reindeer migrating across this region of France some 15,000 years ago (Lewis-Williams, 2002).

These hardy humans were themselves part of a great and long migration. Human life began in the equatorial regions of Africa, where each day is as long as the one before and each season pretty much like the others. But when our great-great-great-grandparents a thousand or more times removed set out from Africa they began to find that there was a seasonal variation in food. Survival at higher latitudes would have depended upon an acute knowledge that animals are not a constant feature of the environment and that offspring are produced only at certain times of year. This knowledge was so important to them that they depicted these migratory events in many cave paintings across Europe.

How does a bison know when it is time to migrate or a salmon know it is time to spawn? The timing is determined largely by the change in photoperiod, or the daily duration of light and dark, which is measured by an internal biological clock. This internal clock is the key to understanding not only when animals start out on these journeys but also how they navigate these annual migrations and the once-in-a-lifetime journeys without charts, compass or wristwatch.

What is true of animals is just as true of plants. Although plants cannot migrate, the annual cycles of planting, growing and the gathering of crops rules the lives of most civilisations. Some plants, such as wheat, flower when the days get longer. Others, such as barley, flower when days shorten. In ancient China, the emperor had to show that his power was a direct extension of the cosmic laws of heaven, and so administration was precisely timed to the days and seasons. Spring was a time of growth and winter a time of death before rebirth. In accordance with these laws of nature, the emperor decreed that winter was the proper time to cut down trees or implement a death sentence (Loewe, 1999).

The priests in ancient Egypt based their power on their ability to predict the timing of the annual flooding of the Nile. They accompanied their promulgations with rituals to give themselves the appearance of supernatural wisdom. In return, the mass of Egyptians living in the Nile delta gave the priests privileged status and a more comfortable existence than that of the common labourer.

All the priests were doing was reading a heavenly calendar. The Nile floods every year when the moon and the sun are in a particular configuration. The ritual and the magic surrounded a vital function. The priests enabled the people to anticipate the flooding. It was essential to know the timing because the flooding did not last very long. As Robin Dunbar has pointed out (Dunbar, 1995):

> *the huge army of labourers had to be raised and in place in time to make the most of the flood's agricultural benefits; they needed some way of predicting the flood long enough in advance to call up a widely dispersed workforce.*

Today, seasonal change has a far less serious impact on the lives of people living in the industrialised nations. In the supermarket there is no summer, winter, spring and autumn – just the commercial contrivances of Christmas puddings and chocolate Easter eggs. The physical implications of the annual cycles that dominated the development of human culture and agriculture are largely diminished in the developed world.

But although we may have obscured the natural rhythm that for thousands of years dictated human life, success still comes, just as it did for the ancient Egyptians, from anticipating the future and organising to meet it. Success in the natural world is a function of living long enough to produce progeny. The more of the progeny that survive to produce progeny in their turn, the greater the measure of success. In the final analysis, reproduction is all that matters, and as Edward O. Wilson put it, 'a chicken really is the chicken genes' way of making more copies of themselves' (Wilson, 1998).

Anticipation is a key to such biological survival and hence success. This is not just the anticipation of a lion chasing a gazelle and guessing which way the gazelle will turn. Nor is it that of a hawk anticipating the movement of a rabbit on the ground so that its dive and the rabbit's run meet at the same point. That is proximal anticipation: behaviour geared to what is appropriate in the immediate future. The anticipation we are talking about is deeper and more profound because it tunes in an organism to its broader environment. François Jacob, one of the great pioneers of molecular biology, said, 'one of the deepest, one of the most general functions of living organisms is to look ahead, to produce future' (Jacob, 1994).

Animals produce future most dramatically by anticipating seasonal processes, and regulate their annual breeding, migration and hibernation to chime with them. Historically much of our understanding of biological clocks has arisen because of our attempts to understand how they do it. In a humbling sense, the scientists who study these events continue a line of enquiry that is older than the Upper Palaeolithic cave paintings in the Grotte de Font-de-Gaume, and may be one of the oldest interests of our species.

Solar radiation is most intense when the sun is directly overhead, and because the earth is more or less spherical there is variation in the intensity of sunlight with latitude. In addition, the earth is tilted 23.5° from its axis of rotation. This means that the amount of energy falling upon our planet will vary during the 365.25 days that it takes for the earth to make one revolution around the sun. In the Northern Hemisphere 21 December marks the beginning of winter, when the north pole is maximally tilted away from the sun. By contrast, the December solstice marks the beginning of the summer in the Southern Hemisphere, when the south pole is maximally tilted towards the sun. The tropics experience the greatest annual input of solar energy and experience the smallest seasonal variation. At the equator the daylength is exactly 12 hours of light and 12 hours of darkness all year round. The seasonal variation in daylength increases progressively towards the poles. The polar regions lie at latitudes greater than 60° north or south of the equator, and experience long cold winters with periods of constant darkness followed by short summers with periods of constant light. At the equator the temperature difference between January and July is less than 1.0°C, at 50°N the difference is 25°C, and at 90°N the difference is 40°C. The amount of solar energy falling upon the earth determines the levels of photosynthesis and hence how productive those regions will be. Food can vary from nothing to massive abundance.

The availability of food around the time of birth is the vital factor determining when young should be born. Lambs are born in the spring, when there is plenty of fresh grass to sustain milk production in the mother and for the young to feed upon after weaning. The type of food that birds feed their young varies between species and determines breeding times. Rooks breed early in the year because the earthworms they feed to their young move deeper into the soil as the warmer, drier days of spring and summer arrive. Finches feed seeds to their chicks, and so produce their young when the grasses have ripened. Eleonora's Falcon breed on islands in the Mediterranean Sea from Greece to Spain. For millennia the species has preyed on songbirds as they migrate over the Mediterranean. The falcons time their breeding activities to the autumn to coincide with the plentiful food supply from the constant stream of birds migrating south to Africa (Follett, 1985).

In the hope of outwitting predators, an additional breeding strategy is to time the production of the young to coincide with that of everyone else. For example, many species breed synchronously and so overwhelm potential predators. The short-tailed shearwater arrives at its breeding site in mid-autumn on small islands north of Tasmania. Essentially, all the individuals of the population lay their eggs between 24 and 27 November each year and hatch at the same time. Such a huge and coordinated production of offspring 'gluts' the predators and allows more young to survive than would occur if offspring emerged over a longer period.

Offspring cannot be produced in short order. Preparation must start well in advance of the burst in food availability. But if an animal does not breed all year around, it is a waste of energy to maintain a fully working set of reproductive organs. Many seasonally breeding animals regress their gonads, which can shrink to almost nothing. In the non-breeding state, the reproductive organs of some seasonally reproductive birds weigh no more than 0.02 per cent of body weight, but in full breeding condition the testes of the male can weigh between 1 and 2 per cent of total body weight (more than the weight of the brain). This is a huge extra load for a bird dependent on flight – the equivalent for a human male of accommodating a one-kilogram bag of flour in his underpants!

It takes about one to two months to go from a fully regressed to a fully working set of reproductive organs. It takes additional time to establish and maintain a breeding territory to attract a mate, and then, after copulation, time for the young to develop before they can emerge into the world. The whole business of procreation can take a couple of months for a small mammal or bird, to over one year in a large mammal. If food were the cue, then by the time the young arrived it would be dwindling or largely gone.

Producing young at the wrong time of year will invariably lead to death. As a result there has been intense evolutionary selection pressure to get seasonal reproduction right. There are potentially many environmental cues that could be used as proximate factors to anticipate the arrival of food and trigger the development of the reproductive organs. But the ultimate factor of evolution by natural selection will favour only the most reliable long-term indicator of the seasons. The change in

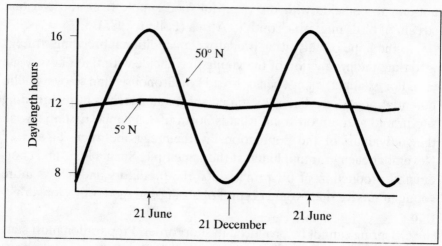

Figure 9.1. Diagram showing the change in daylength with latitude at 50° N and 5° N.

daylength is the calendrical device used by most seasonally breeding animals to adjust not only their annual reproductive cycle but also other seasonal events such as migration and hibernation, and flowering in plants. Figure 9.1 shows how daylength varies at 50°N compared with near the equator. Photoperiodism is the term used to describe such regulation of annual events by the changing daylength.

Farmers and country folk have known about the effects of daylength for centuries and have used it for commercial advantage. More than 200 years ago, Spaniards were providing artificial illumination at night to increase the egg production of chickens (Lofts, 1970), a common practice in the poultry industry of today. Domestic chickens are still very mildly photoperiodic despite the very best efforts to breed chickens that lack this trait.

Cows exposed to light for 16–18 hours a day increase their milk yield in general by 8 to 10 per cent on long days. And nowadays bulb growers in Holland who want to produce year-round tulips drive around on tractors in the middle of the night with big booms of red light that illuminate the plants and fool them into flowering.

Racehorses in the Northern Hemisphere all share 1 January as their official birthday irrespective of their actual birth date. This determines whether they can race as one-year-olds, two-year-olds, or whatever. Consequently, breeders want the mares, which have an 11-month gestation, to conceive in February or March and to foal in January or February. But mares naturally conceive in late spring or early summer and foal in the following late spring. The animals are made to conceive early by keeping the light on in their stalls at night so that they think the days are lengthening.

The most spectacular display of anticipatory timing is the mass migration of birds. Homer likened the routed Trojan army to the fleeing of the cranes from 'the coming winter and sudden rain' (Homer, 1991). In the spring, migratory birds arrive at the breeding grounds in time to stake out a territory, find a mate and then reproduce. In autumn they leave to ensure their winter survival.

The idea that daylength might regulate seasonal breeding was first clearly formulated by E.A. Schäfer in the early 1900s (Schäfer, 1907). But this theory was not tested until the 1920s. William Rowan was a British scientist whose interest in the annual migration of and breeding in birds had taken him to Edmonton, Alberta in Canada. For 14 years he observed the migration of the Greater Yellowleg, a bird that breeds in Canada and then migrates south to Patagonia in the autumn, a round trip of some 16,000 miles. Rowan was struck by the fact that the eggs all hatch between 26 and 29 May each year, and wondered what could regulate this precisely timed series of events (Rowan, 1925).

As with many great insights in biology, unravelling the role of daylength in regulating seasonality started with the detailed observation of a large number of variables. Rowan considered environmental cues such as temperature and food availability, but he concluded that only the change in daylength could provide such precision. He tested this hypothesis by capturing birds called Oregon Juncos (*Junco hymalis*), which overwinter in Canada. Birds were kept in sheds and exposed to artificial spring-like daylengths. Despite the sub-zero temperatures of the Canadian winter, the birds were triggered to breed by the artificially long days (Rowan, 1929).

Rowan's findings have been confirmed in many other temperate bird and mammalian species that have been examined in a similar fashion. All show a photoperiodic reproductive response. The use of artificial photoperiods demonstrated that although secondary factors such as a sudden cold or a warm spell might affect the rate of reproductive development, temperature or food availability by itself could not trigger it. Great tits are heavily dependent upon caterpillars to feed their young, and the number of eggs laid is determined by the availability of caterpillars just before egg-laying time. Caterpillar emergence is determined by environmental temperature, and so temperature indirectly modulates the reproductive potential of these birds, but only by regulating the number of eggs laid – not the timing.

Animals living at the equator, where there is little change in daylength, are far less dependent on photoperiod. The weaver bird (*Quelea*) uses rainfall and the growth of new grass to trigger growth of the reproductive system. Fresh flexible grass stems are of paramount importance to this bird, which builds its remarkably elaborate woven nest from the new-growth grass stems. However, if weaver birds are exposed to artificially long photoperiods, they are stimulated to breed, suggesting that they have retained some capacity to respond to photoperiod (Griffin, 1964).

The changing daylength is used to time reproduction, but the precise daylength selected by a species varies greatly. Birds and small mammals with a relatively short period of development, either in an egg or in the uterus, use the increasing daylengths after 21 December. Large mammals such as sheep and deer have a pregnancy that varies between five and nine months. Breeding is triggered by the decreasing daylengths of autumn. The rut and matings occur between September and December and the young are born in the following spring or early summer. But in some mammals, gestation is even longer: as we have seen, in horses it is 11 months. These long-gestation mammals are stimulated to breed by the increasing daylengths after 21 December. Mating occurs in the spring and summer, with birth nearly a year later.

Although the development time of the young in the uterus (gestation) is fixed in most mammals, some use a neat trick to effectively increase the fixed gestation time so that it synchronises with their mating behaviour. The common harbour seal is usually solitary, the only social cohesion being between the mother and her nursing pup. Although there is a seven- to eight-month gestation, the sexes only meet once a year to mate, which occurs shortly after birth. Delayed implantation allows the female seals to make up the three-month difference, and to give birth and mate at the same time each year. The fertilised egg or blastocyst does not implant straight away in the uterine wall and remains in an arrested developmental state. The blastocyst is entirely dependent on secretions from the wall of the uterus for further development; these secretions are regulated by the ovary, which is in turn regulated by the hypothalamus. Shortening daylengths break this diapause and trigger implantation. All in all, the delay brings the total length of gestation up to 12 months – just in time to give birth and then come back into oestrus and meet a male.

In roe deer, mating occurs in July or August, but the blastocyst implantation is delayed until December, and the fauns are born in May. Without this period of delayed implantation, the young would arrive in December and would stand little chance of survival in the freezing and food-depleted environment at that time of year.

Kangaroos and wallabies have a queuing system known as embryonic diapause whereby the early embryo (rather than the blastocyst) is held in a state of arrested development while a brother or sister occupies the pouch. When the youngster leaves the 'womb with a view', the spare embryo held in the queue will continue development in the real womb. It is born shortly thereafter, and then occupies the recently vacated pouch (Follett, 1985).

At higher latitudes, a difference in daylength of only 8–10 minutes will trigger in many species the reproductive, migratory and hibernatory processes. At the Arctic Circle, this 10-minute change in daylight hours can take place over a two-day period in late summer. The key question is: how does this precise timing come about? One argument likened the measurement of daylength to an hourglass. The idea was that an unknown substance X was converted to substance Y in the light, and then in the dark Y would be converted back to X. When the day was the appropriate length, and there had been enough light falling on the organism, then substance Y reached a critical concentration or threshold and the seasonal event would be triggered.

Erwin Bünning proposed a different mechanism to explain how a circadian clock might be used to time the duration of the night and day and to regulate seasonal events in all organisms. He suggested that what counted was not the amount of light that an animal or plant receives, but critically when it is received. In Bünning's view, there is a rhythm of light sensitivity that is entrained by dawn, and only when light falls at the proper phase or time after dawn will seasonal events be triggered. In other words, there is a specific, critical, photoinducive phase; when this interacts with light, seasonal events are set in train. The difference between this idea and the X>Y hourglass hypothesis is that the animal or

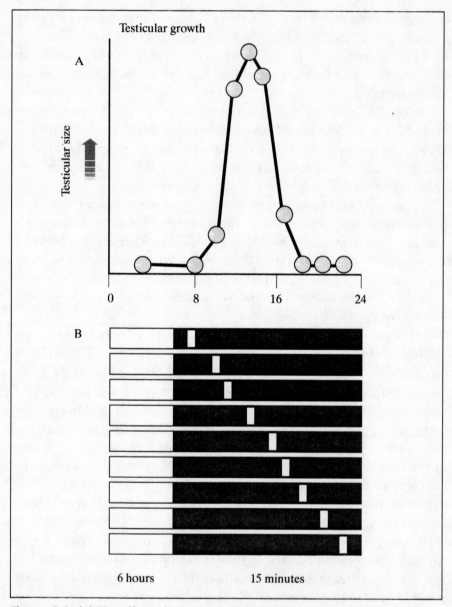

Figure 9.2. (A) The effect of 15-minute light-breaks on the rate of testicular growth in Japanese quail maintained under a six-hour photoperiod. (B) Diagram showing the time of the 15-minute pulse of light. From Follett & Sharp (1969).

plant does not need to actually experience a full 14, 15 or whatever hours of light to trigger reproductive events; it simply needs to experience light at dawn and then light approximately 14, 15 or whatever hours later. The time in between can be spent in darkness.

Distinguishing between a circadian-based and an hourglass-based timing system for photoperiodic responses took the best part of the 1960s and 1970s. Brian Follett and Peter Sharp did one of the key experiments in 1969. They exposed a population of Japanese quail to different 'skeleton photoperiods' (Figure 9.2). All the birds received six hours of light starting at dawn, then different groups of birds were left in darkness before being exposed to 15 minutes of light. For example, some birds were exposed to six hours of light, four hours of darkness and then a 15-minute light pulse, then darkness until dawn, when the cycle was repeated for two weeks. Other groups were exposed to six hours of light, eight hours of darkness and then a 15-minute light pulse, then darkness until dawn, when the cycle was repeated (Figure 9.2).

In each case the birds were only exposed to a total of 6 hours and 15 minutes of light. However, it was only when the 15-minute pulse of light was given between 12 and 16 hours after dawn that the birds were stimulated to breed, as measured by the amount of testicular growth and luteinising hormone (LH) they produced. The birds clearly showed a 'photoinducible phase' or a rhythm of light sensitivity, in line with Bünning's hypothesis (Follett & Sharp, 1969).

The precise position of the photoinducible phase varies from species to species. In the quail the peak of sensitivity is some 14 hours after dawn; in other birds such as house finches it is 12 hours. For most birds, in winter the short days and long nights mean that the photoinducible phase falls during darkness, but as the days lengthen the photoinducible phase will be exposed to light, and seasonal events will be triggered.

That is fine, but what happens to a bird that has a critical daylength in, say, early April? The bird goes into reproductive mode and lays its eggs and hatches them two months later. But with the daylengths still increasing until 21 June, the birds do not want to enter another reproductive cycle, because the young will be born when food supplies are low. What turns reproduction off?

Reproduction in the European starling is triggered by an 11-hour daylength in early spring. But after only another six weeks, and while the daylength is still increasing, the reproductive system collapses. The starling lays its eggs but normally does not copulate again that year. A single-brood species such as the starling is insensitive or 'refractory' to long days, and if kept artificially in long days it will never breed again. The refractoriness is reversed by exposure to the shortening daylengths of autumn, so that the bird will once more respond to the increasing daylengths next spring. Refractoriness in birds is regulated by a complex set of hormonal interactions involving the pituitary hormone prolactin and the thyroid glands. Removal of the thyroid gland blocks reproductive regression by photorefractoriness. In general, photorefractoriness in birds is caused by an insensitivity to a previously stimulatory daylength (Dawson *et al.*, 1986).

In contrast to birds, refractoriness in mammals is the insensitivity to a previously inhibitory daylength. Small mammals such as hamsters will regress their reproductive system in the autumn in direct response to the decreasing daylength. The reproductive system remains regressed over winter but before the days begin to lengthen, the reproductive system starts to spontaneously develop. In this case the hamster is said to have become refractory to the inhibitory effect of short daylengths. Exposure to long days is needed to maintain the animal in breeding and to enable the animal to become sensitive once more to the inhibitory effect of the short days of autumn. In sheep, reproduction is inhibited by the long daylengths of spring and summer, but the animals eventually lose this inhibition and become insensitive or refractory to the inhibitory long daylengths. Reproduction occurs in the autumn, and the increasing daylengths after 21 December trigger reproductive collapse.

Mammals whose SCN has been destroyed can no longer regulate reproduction using changes in daylength. This demonstrates an intimate link between the circadian system and photoperiodic timing. The SCN provides daylength information through its regulation of the pineal hormone melatonin. Russell J. Reiter, Bruce Goldman and others showed that the loss of the pineal, or disconnecting the pineal from its neural connection supplied by the sympathetic nervous system, prevents mammals from distinguishing between long and short days. Removal of the pineal in the golden hamster results in a reproductively active individual irrespective of the photoperiod (Reiter, 1975).

This raises the question of how it all works. What is this somewhat mysterious photoinducible phase? Furthermore, rather than measuring daylength, could the signal that regulates photoperiodic responses actually be night length? It is easy to fool short-day plants into flowering whenever the experimenter wants by exposing them to artificially created short days. But if a dim light is turned on for a short time during the middle of the dark period, the plants are inhibited from flowering. The reverse happens with long-day plants: exposed to short days and long nights they do not flower, but if the long night is interrupted with a brief light pulse then they flower.

If the signal is darkness, or night length, then melatonin is an obvious candidate for the regulatory role. In mammals and birds it is produced in the pineal and released into the blood, where it has a different profile under long and short photoperiods. Under the long nights (short days) of winter, melatonin was found in the blood throughout the night, and animals were exposed to a long-duration melatonin signal. Under the short nights (long days) of summer, the melatonin signal was again found in the blood throughout the night, but of course in this case animals were exposed to a short-duration melatonin signal. Melatonin release from the pineal was co-terminous with the length of the night: long in the winter and short in the spring and summer.

Bruce Goldman and his colleagues removed the pineals from some hamsters. The animals were then kept under constant levels of dim light and divided into two groups. One group received an infusion of melatonin into the blood that corresponded to winter (long-duration infusion), whereas the other group were infused with melatonin that corresponded to spring or summer (short-duration infusion). Those hamsters that received the spring or summer pattern of melatonin were reproductively active, but the winter melatonin pattern did not stimulate reproduction (Goldman et al., 1984). Similar experiments by Fred Karsch and Nancy Wayne in female sheep, which are short-day breeders, showed that in this case it was the winter melatonin pattern that stimulated reproduction, and the spring profile that left the animals unstimulated (Figure 9.3) (Wayne et al., 1988). Rather than light or dark impinging on a photoinducible phase directly, it is the presence or absence of melatonin that signals night length and so regulates the photoperiodic response.

As the SCN lesions block photoperiodic responses, there has to be a method by which the SCN drives this melatonin signal. Electrical activity in the SCN is high during the day but low during the night. This oscillation in electrical activity continues under constant conditions, but

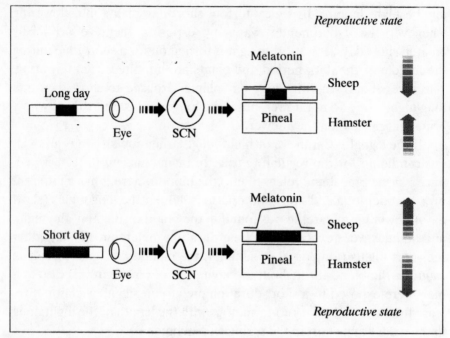

Figure 9.3. Pineal melatonin encodes night length in mammals. The photoperiod is detected by the eye and measured by the circadian clock in the SCN. The SCN drives the synthesis and release of melatonin from the pineal. The duration of the nocturnal release of melatonin decreases as daylength increases. Under long days the duration of nocturnal melatonin is short; this will inhibit short-day breeders such as sheep but stimulate long-day breeders such as hamsters. Under short days the reverse is true.

under a light/dark cycle it is both entrained and modified by light. The SCN signals dusk with a drop in activity, which remains low until dawn; it then increases once more. If a mammal such as a hamster is exposed to either long or short days for several weeks, and then its SCN is removed and its electrical activity monitored in isolation from the rest of the brain, the SCN 'remembers' the daylength. The electrical activity from the SCN mirrors the previous light/dark history even in a dish!

The SCN regulates melatonin release from the pineal by a complicated series of relays through the brain and peripheral nervous system (Figure 9.4). Essentially, the neurotransmitter noradrenaline (also known as norepinephrine) is released from the sympathetic nerve fibres in large amounts during the night. The pattern of noradrenaline

144

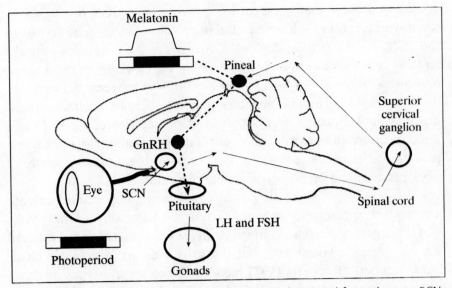

Figure 9.4. Diagram showing the projections to the pineal from the eye, SCN and superior cervical ganglion in a hamster. The duration of melatonin release from the pineal alters the activity of gonadotrophin-releasing hormone (GnRH) neurosecretory cells in the hypothalamus. These project to the pituitary and the release of the neurohormone GnRH in turn stimulates the pituitary to release luteinising hormone (LH) and follicle-stimulating hormone (FSH). These hormones travel in the blood to the reproductive organs (gonads), where they stimulate reproductive activity and the release of testosterone and oestrogen.

release is an inverted image of the electrical activity in the SCN, so that raised electrical activity in the SCN reduces noradrenaline release from sympathetic nerve terminals in the pineal. Adrenergic receptors in pineal cells bind noradrenaline and this ultimately increases calcium levels in the melatonin-producing cells. The calcium signal causes the activation of an enzyme called arylalkylamine *N*-acetyltransferase or AA-NAT. This is the rate-limiting enzyme in melatonin production, and so the rhythmic regulation of AA-NAT by the SCN determines the profile of melatonin production. Furthermore, melatonin feeds back and alters the electrical activity in the SCN so that if melatonin is introduced into the medium bathing a slice of brain containing the SCN, the electrical activity of the SCN is suppressed. In addition, the rhythm of electrical activity of an SCN slice can be phase shifted by melatonin.

We now know that the electrical activity in the SCN driving melatonin synthesis is in itself a reflection of profiles of clock gene activity in the SCN. The longer the daylength, the longer the duration of the *Per* and *Cry* gene expression and vice versa. So the clock genes provide not only a representation of the 24-hour day but also the seasonal variation through the year. The seasonal information does not come from counting individual circadian days but, as we discussed above, results from a direct measure of day or night length. The duration of *Per* and *Cry* expression varies over the year but the period of the circadian profile of clock gene expression is approximately 24 hours.

The details of how the melatonin signal itself is decoded to generate a reproductive response are complicated, not helped by a terminology that takes getting used to. Unfortunately, at times there is no shorthand for the scientific vocabulary, but the principles are reasonably clear. Steve Reppert (Reppert, 1997) has discovered several types of melatonin receptors in all mammals, including humans, both in the SCN and in a small area of the pituitary gland called the pars tuberalis. In an as yet unknown manner, melatonin acts on the SCN and the areas around the SCN to regulate the output signal to one of the primary targets for photoperiodic information, a group of cells in the hypothalamus called gonadotrophin-releasing hormone (GnRH) neurosecretory cells. These project to the pituitary gland, and the release of the neurohormone GnRH in turn stimulates the pituitary to release luteinizing hormone (LH) and follicle-stimulating hormone (FSH). These hormones travel in the blood to the reproductive organs, where they stimulate reproductive activity and the release of testosterone and oestrogen (Figure 9.4).

The pattern of GnRH release is critical. The neurohormone does not dribble out of the end of GnRH neurones in a continuous stream but is released in pulses. The pulse interval is decoded by the pituitary, and this alters the pattern of release of LH and FSH. In sheep (short-day breeders) there is a high pulse frequency of GnRH and then LH under short days and this ultimately triggers the reproductive system. But under long days the pulse frequency of GnRH and consequently LH is long, so the pituitary is not stimulated to release the reproductive hormones.

The pars tuberalis, which is rich in melatonin receptors, regulates the production of the hormone prolactin. In the hamster, a long day (short melatonin profile) causes high levels of prolactin secretion, whereas short days (long melatonin profile) suppress prolactin secretion. Clock genes in the pars tuberalis regulate cellular activity. The melatonin signal controls the local pattern of expression of *Pers* and *Crys*, which in turn modulates the sequence that leads to the synthesis and release of prolactin. Prolactin signals the mammary glands to produce milk. It is also involved in regulating other aspects of seasonal physiology, including food intake, changes in metabolic rate, and winter coat growth and colour (Lincoln, 1999).

Given that day length (or night length) causes an altered pattern of clock gene expression in both the SCN and the pars tuberalis, and that these changes have been associated with the regulation of seasonal physiology, it was thought that the same mechanism might explain refractory insensitivity. The hamster's seasonal reproductive response is 'turned off' by the short days of autumn. Kept on short days for about three months (winter time), the inhibitory effect disappears and the hamster spontaneously 'switches on' its reproductive response.

As the melatonin signal throughout this period continues to mirror day length (or night length), and the localised pattern of clock gene expression in the pars tuberalis parallels the melatonin signal, it is not the melatonin signal itself that triggers the refractory response. The melatonin signal must be read differently by target tissues at different times of the year. But how this happens is completely unknown.

The story is not yet complete, but the melatonin profile released from the pineal that mirrors night length is the critical cue that drives seasonal events in mammals. Although humans share all the basic anatomical and molecular components of photoperiodic mammals, and some of our primate relatives show a robust photoperiodic response, nowadays there seems to be only a subtle seasonal variation in human reproduction. This has been attributed to many factors, one of them being an increase in temperature during the summer resulting in decreased sperm production and fertility in males. So much for global warming!

Melatonin has a crucial role in the timing of reproductive activity. Reproduction has system-wide effects on an organism. The melatonin signal has been selected by evolutionary process in mammals for reproductive control. The use of melatonin for obviating symptoms of jet-lag and for regulating sleep may well have minimal and acceptable side-effects. However, there is a precautionary view that a substance with

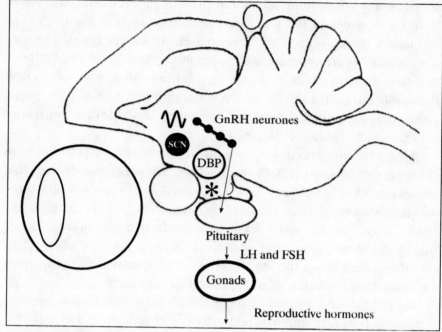

Figure 9.5. The organisation of the photoperiodic machinery in a bird. Daylength is detected by deep-brain photoreceptors (DBP) and measured by a circadian clock in the SCN. Daylength information then regulates the activity of gonadotrophin-releasing hormone (GnRH) neurones in the hypothalamus. This stimulates the pituitary gland to release the hormones LH (luteinising hormone) and FSH (follicle-stimulating hormone). These travel in the blood and reach the gonads. In the male the testis is stimulated to produce testosterone, and in the female the ovary produces oestrogen. The pineal and eye, so important in the circadian system of many birds, are not required for the photoperiodic regulation of reproduction.

such profound effects on the reproductive system and general metabolism should be treated with some care.

The organisation of the circadian system of birds and mammals differs markedly, and this is also true for the components regulating the photoperiodic response of these two groups. The pineal of most birds contains a self-sustained circadian oscillator that regulates the release of melatonin from the gland. The pineal can be removed and maintained in culture medium in a dish, and the pattern of melatonin release is determined by the length of the dark portion of a light/dark cycle. Furthermore, when the gland is placed in darkness the pattern of melatonin release 'remembers' the pattern of release under the previous light/dark cycle. Like the mammalian pineal, the avian pineal can code for the length of the night in terms of melatonin release.

But unlike in mammals, removal of the pineal has little or no effect on the ability of birds to show seasonal reproductive cycles! The melatonin signal of birds is not used as an indicator of night length for reproduction. Furthermore, the photoperiodic clock in birds is in the hypothalamus. Whereas in the circadian system the entraining light signal is provided by a combination of eyes, pineal and deep-brain photoreceptors, the light signal for photoperiodic time measurement comes exclusively from the deep-brain photoreceptors (Figure 9.5).

Jacques Benoit had shown in the 1930s that blinding mallard ducks did not block the photoperiodic response. This has now been shown to be true for all birds. The deep-brain photoreceptors were finally located in the 1980s by using fine fibre optics to illuminate small regions of the brain, thus providing a long day to a specific region. Two brain areas were identified, one in the hypothalamus and a second region in front of the hypothalamus. These photoreceptors use an opsin-based photopigment, which is located in a group of cells called cerebrospinal fluid (CSF)-contacting neurones. These deep-brain photoreceptors are thought to regulate exclusively the photoperiodic response and contribute to circadian responses to light in birds (Foster & Soni, 1998).

In birds, deep-brain photoreceptors detect the photoperiod, and a circadian clock in the SCN measures daylength. In some as yet unknown way this daylength signal is then used to drive photoperiodic responses to light. In terms of the reproductive response, as in mammals, the pattern of release of GnRH from a dedicated set of neurosecretory neurones alters the secretion of a number of hormones from the pituitary gland that in turn stimulate the gonads and the production of the sex steroid hormones (Figure 9.5).

There are other photoperiodically regulated events in birds, such as migration, the moulting of feathers and the regulation of song. Birds must time their entire annual cycle so that the arrival of the young in the nest coincides with an optimal abundance of food, and they may need to migrate seasonally to find these conditions. Migration has been better described than understood. The actual migration is quite a performance, but it comes at the end of several weeks' build-up. Migrating birds can double their body weight, laying down stores of fat that are used up by the end of the journey. How much fat they store depends on the distance that needs to be travelled. A typical blackpoll warbler at the end of its breeding season weighs about 11 grams. In preparing for its transatlantic trek, it may accumulate enough fat reserves to increase its body weight to 21 grams. Given an in-flight fat consumption rate of 0.6 per cent of its body weight per hour, the bird then has enough added fuel for approximately 90 hours of flight for a journey that, under fair conditions, requires about 80–90 hours. The 14 grams of fat in a single Snickers bar would provide one and a half times the amount of energy necessary for the Blackpoll Warbler's flight from New England to South America. If the warbler were burning petrol instead of reserves of body fat, it would be getting 720,000 miles to the gallon (Deinlein, 1997).

Most birds are diurnal, yet many, including most shorebirds and songbirds, radically shift to night-time activity when they migrate, overriding their normal circadian behaviour. The possible advantages to flying at night include decreased vulnerability to predators, reduced threat of dehydration or overheating, a greater likelihood of encountering favourable winds and a stable air mass (rising hot air and more variable wind directions occur during the daytime), and time during the day to forage.

Night-time migration is mimicked in captivity in a process known as *Zugunruhe*. This occurs twice a year during spring: when birds migrate to the breeding grounds, and then again in autumn, when birds move back from the breeding to the wintering areas. Caged birds hop around in the general direction they would head off to if they were migrating. Although it was well known to bird fanciers, it was first studied properly in 1949 by Gustav Kramer. He designed and built a circular cage with a circular perch. The observer had to lie under the glass floor of the cage and observe which side of the cage migratorily active birds preferred at night. This behaviour, characterised by rapid fluttering of the wings while perching, begins at the same time that conspecifics (individuals of the same species) in the wild are setting off on migration, and persists for the same length of time required for the wild counterparts to complete their migration (Kramer, 1949, 1952). *Zugunruhe* has been used as a key research tool to study bird migration in the laboratory, although it has seldom been reported in the wild. Non-migratory birds do not exhibit this behaviour.

Despite the intense interest in migration by both professional scientists and thousands of amateur bird watchers, we have a very limited understanding of the physiological mechanisms that trigger these journeys. The lengthening day seems to be a key trigger in spring migration to sub-Arctic regions by birds wintering in northern temperate zones (such as the UK or North America). It was assumed that the physiological signal for this was the rise in the levels of reproductive hormones. This turned out not to be the case, because removing the testes reduces these hormone levels but does not block male migration in most birds. Another assumption was that the fat reserves laid down before migration might in some way signal *Zugunruhe*. However, experimentally blocking the laying down of stored fat does not stop nocturnal restlessness. There must be some independent photoperiodically driven signal that drives *Zugunruhe*, but we do not know what it is.

As soon as the young have been born and the initial frantic feeding is over, migratory birds start preparing for the return flight. In any year, birds spend more than half their time preparing to migrate and actually migrating. Far less is known about the triggers for the return journey in autumn. It may be the decrease in daylength, rainfall, a drop in environmental temperature, or the fact that the birds are photorefractory. Some species may have a separate clock that triggers this event.

Plumage renewal or the moult is also triggered by a photoperiodic response. It is species dependent, but usually occurs after the breeding season when the young need little or no care. During this time the entire plumage, including flight and tail feathers, are completely replaced as damaged feathers need to be renewed. Fat is also deposited and stored in the body.

Timing of the moult is a particular problem for migratory birds breeding at high latitudes, as plumage renewal has to be completed before conditions become life threatening. If the birds waited until the climate became no longer tolerable to begin preparations for leaving the breeding areas, it would be too late to gain the necessary energy surplus to allow the required physiological changes associated with migration. The Arctic tern, which migrates between the Antarctic and the Arctic, delays its moult until it has returned to its southern base, and some species such as barn swallows moult while they are migrating.

Birds sing mainly in spring and the male does most of the singing, again triggered by changes in daylength. For centuries, Japanese bird fanciers have practised the art of yogai, in which cage birds are forced to sing in midwinter, by lengthening their days using candlelight for three or four hours per day after sunset. Dutch hunters similarly stimulated various finches to sing prematurely in October for the purpose of enticing autumn migrating finches into traps.

The male bird sings to announce that he is in residence, and uses song to establish the limits of his territory and to advertise for a mate. The current explanation of this seasonality is that the nuclei in the bird brain involved in the production and learning of song increase in volume during the breeding season and then shrink again out of breeding. Part of this change is regulated by hormones from the reproductive organs. However, if melatonin is given to birds on long days then the normal increase in song nuclei is attenuated, suggesting that in this special case the melatonin signal from the pineal is somehow involved in regulating the singing behaviour.

Much of the thinking about the timing of annual events has been based upon our growing understanding of the measurement of daylength by the circadian system. There is, however, an alternative mechanism that can time annual events through the use of a separate clock.

There is a practical difficulty in researching annual events because by definition the experiments take years, unlike molecular biology, in which you can get results in an afternoon. The golden-mantled ground squirrel (*Citellus lateralis*) naturally hibernates. In one experiment, some squirrels were brought into the lab and maintained in constant-temperature rooms set permanently at 3°C and exposed to three different light cycles of either constant dark (DD), constant light (LL), or 12 hours light and 12 hours dark (LD 12:12). In all these conditions the squirrels hibernated each year. The time between hibernations was around 11 months rather than exactly one year, and the individual squirrels had slightly different periods. Hibernation had to be timed by some endogenous clock as it was not being driven by some unrecognised environmental event within the laboratory. Even golden-mantled ground squirrels born in captivity and who have never experienced a natural photoperiod still have a circannual rhythm (Pengelley & Fisher, 1966).

Studying these rhythms requires an enormous amount of patience. It took Pengelley and Fisher five years to get their squirrel results. Eberhard Gwinner has been working on circannual clocks for over 30 years and some of his experiments cover 12-year periods. In the early 1970s, he and his team showed that the garden warbler (*Sylvia borin*) had circannual rhythms. The bird is a long-distance migrant that displays intense *Zugunruhe* during the migratory season in spring and autumn. Like the studies on ground squirrels, warblers were brought into controlled temperature and light facilities and maintained under LD 10:14, 12:12 and 16:8 hours. Rhythms of *Zugunruhe*, body weight and moult continued for many years, and again like the ground squirrels the period of these rhythms deviated significantly from 12 months (Gwinner, 1996a). Later experiments by Gwinner and colleagues suggested that there might be several independent circannual clocks, because simultaneously recorded rhythms in testicular size, moult, body weight and *Zugunruhe* free-run with different periods.

Under normal circumstances the circannual rhythms of many species are entrained by the changes in daylength. However, some circannual species such as the stonechat (*Saxicola torquata axillaris*) live and breed at the equator, where the daylength is 12 hours the whole year round. The question that Gwinner has been trying to address is: how do these species entrain their circannual rhythms? One possibility was food availability, and so Gwinner brought stonechats into the laboratory under controlled LD 12:12 conditions, and provided extra food two months in advance of the normal breeding season. He reasoned that if food availability could act as a *Zeitgeber* then one would expect breeding to occur two months earlier the following season. It does not, so Gwinner is looking at other potential entraining agents, such as rainfall and light quality (Scheuerlein & Gwinner, 2002).

Over 30 different species have circannual rhythms, including unicellular organisms, invertebrates such as molluscs (slugs) and arthropods (beetles), as well as the different groups of vertebrates. These annual clocks may be fairly common in long-lived species. But to make an already cloudy picture even murkier, some species will only express circannual rhythms under certain circumstances. European starlings have circannual rhythms of testis growth and regression under LD conditions of 12:12, but not when exposed to shorter or longer photoperiods. The starlings, of course, become photorefractory under long daylengths, and will not breed at all in shorter photoperiods of less than 11 hours of light. The circadian system would seem in this case to override the circannual response.

There is a very complex interaction between circadian and circannual clocks in the regulation of many seasonal events. Sheep have been one of the classical models for studying the role of a circadian clock in photoperiodism. If sheep are maintained under a long day (16:8 LD) cycle then they too will show a circannual cycle in reproduction but not under a short day cycle. It seems that the circadian-based photoperiodic regulation of reproduction is simply one mechanism for the timing of reproductive effort.

The existence of an independent and truly annual clock mechanism was contentious with some, who argued that circannual rhythms might in some way be derived from circadian oscillators. For example, in the same way that an atomic clock counts the high-frequency oscillations of the caesium atom and turns them into hours and minutes, perhaps circannual clocks count circadian frequencies and transform them into circannual timers. If circannual clocks were simply devices that counted 365 circadian days, then the number of circadian days should determine the period of the circannual rhythm. To test this possibility, several bird species have been maintained on either LD 11:11 or 12:12 (either a 22-hour or a 24-hour day) for almost three years. The period of the circannual rhythm in each case did not differ significantly, suggesting that circannual rhythms do not arise by counting approximately 365 circadian cycles.

Where the circannual clocks reside in birds and mammals, or indeed whether there is an anatomical localisation for the circannual mechanism, remains an open question. It is certainly distinct from the circadian system. SCN lesions in golden-mantled ground squirrels eliminate circadian rhythms of behaviour but do not abolish circannual body-weight cycles or hibernation. Although there is some disruption to these cycles, the SCN cannot be an essential component of the circannual mechanism.

Organisms living near the equator are in a near-constant daily light/darkness cycle. Those that live at high Arctic latitudes are in continuous darkness between November and February, and continuous light from April to September. The continuous light of the polar day and dark of the polar night represent a special problem for the circadian system, which uses light as its primary *Zeitgeber*. Three patterns of circadian behaviour have been recorded in different species under these conditions. Migratory birds visiting the high Arctic in the summer to breed show entrained circadian activity, suggesting that the subtle change in the light environment over 24 hours provides a sufficiently strong *Zeitgeber* for them. In the same environment, humans working on polar bases are unable to entrain and show free-running rhythms of sleep and wake. Surprisingly, native birds and mammals such as ptarmigan and reindeer do not entrain to the local condition and become arrhythmic, showing ultradian bouts of activity with no obvious circadian component. These species also show a flattened melatonin profile over the same period (Reierth *et al.*, 1999).

Karl Arne Stokkan, at the Department of Arctic Biology at the University of Tromsø, has suggested that this lack of circadian rhythmicity may be an important adaptation to this extreme environment. 'At those times of the year when the environment becomes arrhythmic and unpredictable, the circadian system is flexible and increases fitness by allowing these herbivorous animals to forage whenever physical conditions are favourable.' What Stokkan means is that in the unpredictable environment of the high Arctic the native animals are released from circadian constraints and can feed opportunistically (Reierth & Stokkan, 2002).

Flattened melatonin profiles have also been observed in birds during migration. The amplitude of the plasma melatonin rhythm in captive garden warblers is much reduced during the migratory season in comparison with other times of the year. Gwinner believes that birds that migrate rapidly along an east–west axis, and cross multiple longitudes and hence time zones, need to be able to resynchronise rapidly to the changing environmental *Zeitgeber* to prevent avian 'jet-lag'. Rapid migration along a north–south axis will also require entrainment to a changing daylength. The reduced amplitude of the melatonin signal is likely to reduce the robustness or self-sustainment of the circadian system. Such a system will be able to resynchronise faster to the changing dawn/dusk signal. In effect, a weakened oscillatory rhythm will be easier to push about by light than a robust and strongly self-sustained oscillator (Brandstatter, 2002).

This raises the question of why a reduction in self-sustainment would not be an advantage all the time. The argument here is that in non-migratory species, or when a migratory species is not migrating, the precision of entrainment is what counts. You do not want the oscillator to be pushed around easily by minor 'noisy' fluctuations in the light environment. A high-amplitude, highly self-sustained system due to a strong melatonin cycle will counteract environmental 'noise' and allow very precise entrainment. This leads to a prediction about the amplitude of melatonin release in migratory birds that visit the high Arctic in the summer and that do entrain. The *Zeitgeber* signal is very weak, and a weak oscillator is easier to entrain by a noisy signal than a strongly self-sustained oscillator. So it follows that migratory birds visiting the Arctic during the summer would have low-amplitude melatonin rhythms and weakly sustained oscillators. And they do.

Migratory birds are unique in that individuals cover huge areas of the planet on their travels and they live out their lives in wildly fluctuating temporal environments. It should not be too surprising that not only do they have the most sophisticated endogenous timing devices, but the wide variation in the conditions they face means that there are inter-specific differences. But the wider point is that the timing of seasonal reproduction, and the seasonal events linked to it, provides a spectacular example of the importance of an endogenous timing system that can predict environmental change. Optimising the production of young to the appropriate time of the year will be under the most intense selection pressure. The mechanisms are complex and involve not just one type of endogenous clock but two. Circadian and circannual timing systems interact to provide a dynamic predictive programme that can adapt to the changing requirements of the organism and to an environment with ever-changing *Zeitgeber* signals.

10

CLOCKWORK EVOLUTION

*What mole rats have given me is a reawakening of wonder, in many ways.
For me as a biologist it's been the mole rats, reminding me that there's a
whole world out there that continuously has to be re-explored and continuously has to be questioned; that all that we know only relates to a moment in time, cultural views, scientific ethics, and positions held by people
in power. But the questioning has nothing to do with that. The questioning
has to do with the purest form of wonder. 'Look.'*

RAY MENDEZ IN *FAST, CHEAP AND OUT OF CONTROL –*
A FILM BY ERROL MORRIS (1997)

There is a scene in the film *Silence of the Lambs* when they find death's
head moths on a corpse. Ray Mendez painted those moths so they would
show up better on screen. He did in a few minutes something that would
take evolution a few million years. But that is Hollywood for you.

Mendez's day job is as an insect wrangler, working with film, media
and zoo people who need insects. Apart from the moths, he provided the
somnolent honeybees for an IBM advertisement. But his passion is a
mammal – the naked mole rat. He kept a colony of the small, hairless
burrowing creatures in his apartment when he lived in New York.

At first, second and third glance it is hard to see why he is so keen on
them. Naked mole rats have to be the world's least attractive animals.
Although they are related to cuddly guinea-pigs, these subterranean
creatures are only about 10–15 centimetres long with tiny eyes and no

ears. They have almost no body hair and no fat beneath their skin, which gives them a wrinkly, aged appearance. They have the poorest capacity for thermoregulation of any mammal, but they thrive in subterranean burrows whose stable climate helps them maintain their body temperature of 31°C. With their pink, wrinkled, hairless bodies they resemble chipolata sausages. And then there are the teeth. The extra-long incisors extend in front of the lips like a prosthetic claw. The rats use their protruding teeth to bite through the earth without getting it in their mouths. The result is that they resemble, in the words of one anonymous Australian researcher, 'dicks with teeth'.

Those teeth can bite through anything. Mendez explains:

> we were designing enclosures to put the naked mole rats in but they have incredible teeth. Over 10% of their body weight is dedicated to their jaw muscles. And so when we put the animals together in a chamber, that would look like it was underground, constructed underground by them, they would chew right through it. They could chew through plastic. So I built chambers out of concrete and they chewed their way through concrete. Which brought about many science fiction possibilities. I mean, you know, the ultimate bug that escaped in New York and all the skyscrapers that come down.

Naked mole rats are not natives of Manhattan. They live in colonies of 70 or more in an intricate tunnel system that can cover several kilometres beneath the arid earth of eastern Africa. Life is not easy. The naked mole rats spend most of their time digging through the hard earth searching for basketball-sized plant tubers that are full of the essential water they need. A solitary naked mole rat would not survive long. Instead they work in teams, gnawing at the earth with their outsize incisors and shovelling the earth back along the line with their feet.

Apart from their cousins the Damaraland mole rat, naked mole rats are the only eusocial (truly social, from *eu*, the Greek for true) mammals that we know of. But unlike their insect counterparts, naked mole rats are not irrevocably fixed in their caste. All are capable of developing to sexual maturity.

Apart from this eusociality, one of the most interesting questions is how this closely knit community forms the new colonies that are essential for genetic variation. Sometimes, an animal leaves the colony and moves over large distances on the surface. One marked female was found two kilometres from her colony. We can assume that many do not make it, because they are such an easy target for owls and other hunters. But some obviously do, meet a mate and form a new, successful colony.

They time this new colony formation, just as the ants and termites do, for the end of the rainy season, presumably when there is lots of food and the soil is soft. Other mole rat species in South Africa go a stage further and have dispersion morphs, which are bigger than the average animal. The dispersion morph has grown in anticipation of when it will leave to start a new colony. But mole rats have to have a sense of timing if they are going to maximise the chances of successful new colony formation. A naked mole rat lives deep underground. It is more than somewhat impervious to the external environment and the change between night and day. Yet a naked mole rat has a circadian rhythm entrained to the 24-hour cycle and it can time the migration to find a mate.

Presumably the naked mole rat has retained a circadian system from its ancestry and it can at least sense light on the occasions when it surfaces to clear detritus. Its cousin, the solitary-living blind mole rat, however, has no functioning visual apparatus. *Spalax ehrenbergi* is a subterranean rodent with atrophied eyes. Those regions of the brain associated with the image-forming visual system have been reduced by a massive 87–97 per cent. Although visually blind it has minute eyes of little more than 0.5 millimetres in diameter buried under the skin.

Yet *Spalax* also can use light for photoentrainment. Over the past 30 million years, evolutionary processes seem to have disentangled and eliminated the image-forming visual system of this animal while retaining those components of the eye that regulate the biological clock. The retinohypothalamic tract is intact even if the visual processing centres in the cortex are not there.

This raises the question, why bother? Nature is seldom profligate. If something is not needed, particularly anything connected to the brain that uses up energy, then it disappears. The mole rat lives in a virtually constant-temperature, constantly dark underground environment. It even meets a mate underground. Individual blind mole rats signal a willingness to mate by banging their heads on their tunnel roof. This vibration is picked up by a neighbour, and the separating tunnel walls are broken down, after which mating occurs. Despite this largely subterranean existence, *Spalax* seems to need its circadian rhythm, and the animals are clearly rhythmic in their metabolism and behaviour. Presumably, the ability to anticipate dawn and/or dusk still provides some selective advantage. Or perhaps blind mole rats just need some temporal mechanism to maintain their overall metabolism in coherent order.

Presumptions, however, are not evidence. Carl Hirschie Johnson of Vanderbilt University, Nashville, Tennessee, who is one of the leading investigators in the evolution of circadian clocks, has commented, 'Temporal programming by circadian oscillators must be important, otherwise why should such an elaborate and pervasive regulatory system have evolved?' Why indeed? The answer ought to be easy. Endogenous circadian rhythms are adaptive, as they help to improve both the survival of an organism and – critically, and however marginally – its success at reproducing itself. But saying it is not the same as proving it. As Johnson points out, 'Nevertheless, it is surprising how little experimental support we have for the value in terms of the reproductive fitness of circadian rhythms' (Johnson *et al.*, 1998).

In principle we will never know fully how the circadian system evolved, but the beauty of the theory of evolution is its explanatory power. Darwin's simple axioms provided the basis for a materialist account of the development of life from inorganic molecules through to the diversity of organisms alive today. Thinking about the evolution of circadian rhythms has been, in Michael Menaker's words, 'of real heuristic value to the field' (Menaker & Tosini, 1996).

Reproductive advantage does not have to be large for a feature or behaviour to be adaptive. Many butterflies have eyespots, but the Brazilian butterfly *Caligo* goes furthest in having eyespots that closely mimic a vertebrate eye. *Caligo*'s markings mimic large bulging vertebrate eyes set in deep sockets. There are even some flecks of white set in an arc, representing highlights reflected off the glistening eye surface. Manning and Stamp-Dawkins (1998) have explained this seemingly excessive attention to detail when they write:

> *because changing the pigment of those scales on its wings that go to make up the eye-spots has presumably only trivial costs for* Caligo, *there has been no barrier to selection operating over hundreds of thousands of generations to bring the eye to ever more perfect levels of adaptiveness. Each tiny advance will have brought only tiny advantages, but there has been*

time for even the tiniest advantage to pay off in terms of slightly more effec-
tive bird scaring and hence result in the survival of a few more of those but-
terflies which bear it.

Time is the one thing that evolution has had plenty of.

Several plant species, including tomato, grow very poorly in the absence of environmental time cues, or in light/dark cycles that differ markedly from 24 hours, indicating that some type of timing in the circadian range is required for normal growth (Highkin & Hanson, 1954). In the 1950s, Fritz Went and his students at the Cal. Tech. Phytotron found that different light/dark cycle lengths affected the growth of African violets (*Saintpaulia*). Growth varied markedly and was low on the longer and shorter periods and maximal on the 24-hour cycle. The key finding was that it was the period itself that mattered, but Went's studies were hardly conclusive evidence that circadian rhythms were adaptive (Went, 1960).

Despite years of study, the laboratory evidence for circadian rhythms in rodents being adaptive is barely encouraging. Mammals that have had their SCN removed in the laboratory survive when kept in these comfortable conditions. Golden-mantled squirrels caught in the wild survived for more than two and a half years in the laboratory after lesioning.

Such model animals are generally chosen for ease of handling, breeding, and manipulating various characters, rather than for their natural history. A long-standing criticism of evolutionary theory's 'modern synthesis' is that it has focused on micromutations at the expense of adequately considering the role of ecological pressures. 'Model organism research asks fundamental questions about how the animal works', comments David N. Reznick, a biology professor at University of California, Riverside. 'It isn't done with regard to how the animal functions in nature' (Bunk, 2002).

Circadian researchers suffer a double whammy in that apart from using model animals, they also use highly artificial environmental conditions for their experiments. Till Roenneberg of Munich University says (personal communication):

Most experiments are conducted under constant conditions, which are highly
unlikely in nature. However, such conditions are necessary to investigate

endogenous rhythms and their responses to stimuli. But how constant can the laboratory environment be when so many functions of an organism from enzyme activity to behaviour oscillate? For example Gonyaulax cells are kept in small volumes of medium on which the cells impose systematic changes due to their circadian physiology. Throughout the cycle they take up nitrate at different rates and the normal rate of cell death within a culture makes more nitrate available mainly at the end of the subjective night. Thus the endogenous system creates a nitrate oscillation in the exogenous medium.

Patricia DeCoursey of the University of South Carolina has worked on circadian rhythms in mammals for over 40 years. She has tried to answer the question of adaptiveness of entrainable circadian rhythms in the wild. It is difficult, time-consuming and not always rewarding work. Animals outside the comforts of the laboratory have to contend with predators, climate and the search for food. And it is not much fun for the researchers who work long shifts in often inclement weather (DeCoursey *et al.*, 2000)

In a pilot study, DeCoursey kept antelope ground squirrels in an enclosure that mimicked natural desert conditions. A feral cat opportunistically played the role of natural selection when it managed to get into the enclosure. 'Continuous infra-red video taping showed graphically that the unsuspecting SCN-lesioned individuals wandered out of their burrows at night without adequate night vision and were killed immediately by the waiting cat.' This extreme selection pressure was promptly attenuated. She shot the cat (DeCoursey *et al.*, 1997).

The main study was a massive 18-month experiment in the Allegheny mountains of Virginia. She and her students racked up 4,200 hours studying what happened to chipmunks whose SCN had been destroyed by lesioning. In all, 74 animals were monitored on the four-hectare site. Thirty animals were SCN lesioned, another 24 were operated upon but the SCN was not destroyed, and there were a further 20 intact controls. The chipmunks had to contend with hawks and rattlesnakes, but, dangerous as they can be to chipmunks, in this case the agent of natural selection was the weasel. Weasels can eat their own weight in prey each day. They sever the spinal cord of the prey at the cervical vertebrae, cut into the dorsal brain case and eat the brain. But unlike the initially lucky feral cat in the compound, the weasel did not pick up its prey opportunistically at night on the surface. The chipmunks stayed in the comparative safety of their dens.

However, some clue allowed the predator to target the chipmunks that had been operated on. Whatever the reason, the weasel got them. In the first 14 days of the experiment, 37.5 per cent of the surgical controls and 50 per cent of the lesioned chipmunks died, while all the non-surgical controls survived. In her pilot study, there had been little mortality in the first few days and DeCoursey put the mayhem in the main study down to the weather.

> Chipmunk populations of the Alleghenies are particularly susceptible to oak productivity cycles. A cyclic rise in chipmunk population density due to two consecutive highly favourable acorn crops in 1995 and 1996 favoured reproduction during the pilot study and the chipmunk population increased explosively. By summer 1997, chipmunks had reached saturation density. The abundance of chipmunks in turn attracted a rapacious carnivore, which captured predominantly SCN-lesioned chipmunks as prey. Within a few weeks the lesioned chipmunks were decimated in numbers.

Which was kind of tough on the lesioned animals (DeCoursey *et al.*, 2000).

After the carnage of the first fortnight, the pattern of predation settled down, and over the course of the next 66 days 30 per cent of the controls were killed, as were 20 per cent of the surviving surgical shams and 66.7 per cent of the SCN-lesioned animals. DeCoursey *et al.* (2000) concluded:

> When climatic and biological factors were highly favourable in the chipmunk population, the dependence on the SCN did not seem very great. As times became less ideal from overcrowding and especially from the upswing of the predator population, decimation of the SCN-lesioned segment of the chipmunk population was stunningly evident.

So, after 4,200 hours of work there seemed to be reasonable evidence that rhythmic animals have some survival advantage. But useful as it is for animals to survive, in itself it says little about the evolutionary value. Darwin's theory is not interested in the animals themselves, but in their reproductive success. Natural selection works on the whole organism, but evolutionary advantage comes from the propagation of the genes. As long as the animal reproduces and if necessary nurtures the young to independence, then what happens to it afterwards is of no value to nature – animals, including us, were not meant to be old. Richard Dawkins, in one of his hyperbolic passages, put it like this (Dawkins, 1976):

> a gene does not grow senile; it is no more likely to die when it is a million years old than when it is only a hundred. It leaps from body to body down the generations, manipulating body after body in its own way for its own ends, abandoning a succession of mortal bodies before they sink in senility and death.

The chipmunks and any offspring would have had to be monitored for at least the following year – another 4,200 hours of work – to show that circadian rhythms are adaptive. And if climactic factors intervened it might have needed a further season after that, and who knows how many more seasons, to put the facts of evolutionary advantage beyond question.

No wonder biologists prefer to work with bugs. They are born, live and die within the space of hours if not minutes; 4,200 hours buys a lot of research into bacteria. They are divided into two main groups: the Archaebacteria – the best known of which live in the high-temperature, high-acidity settings of the hydrothermal vents deep under the sea – which live by converting into usable energy what to them is a rich range of nutrients such as hydrogen sulphide; and the Eubacteria, with which we are familiar – they include the pathogens that transmit cholera and the beneficial bacterial flora in our intestines.

Neither Eubacteria nor Archaebacteria have a nucleus, and they are known as prokaryotes (before nucleus). They were both around well before eukaryotes (truly nucleated organisms). The first eukaryotes probably did not appear until about two billion years ago.

Without bacteria we would not be here to observe them. We owe our existence to the production of an oxygen-rich atmosphere largely created by the efforts of various strains of bacteria photosynthesising for several billion years. To put it bluntly, we live in bacterial excreta. It is thought that the first cyanobacteria appeared more than 3.5 billion years ago when the earth's atmosphere was very different from today, though it has to be said that determining whether a tiny speck of carbon in an old rock is due to life processes or more prosaic volcanic events is fiendishly difficult. But give or take a few hundred million years, cyanobacteria have been around for a long time. The environment in those early days has been well described by William Schopf (Schopf, 1999):

> we can surmise that whatever continents that existed were small by present-day standards and that they trucked around the global surface faster than today, powered by the large amounts of heat escaping from the Earth's interior. The Earth revolved more rapidly and days were shorter, tides greater, storms more severe. The skies were a hazy steel-grey blue, darkened by dust storms, volcanic clouds and fine rocky debris kicked up by bombarding meteorites. The atmosphere was rich in nitrogen, carbon dioxide and water vapour but contained only trace amounts of oxygen gas [the oxygen level was 10^{-13} bars]. Because of the near absence of free oxygen, ultra-violet absorbing ozone was in short supply, so the Earth's surface was bathed in ultra-violet light lethal to early life.

Cyanobacteria are found today living on the surfaces of ponds and as large layered structures known as stromatolites in shallow waters. These mats of cyanobacteria grow in an aquatic environment, trapping sediment and sometimes secreting calcium carbonate. Over three billion years ago stromatolites were a common sight. (It is very hard to come to terms with such long periods of time, but the key to understanding evolution is the realisation of just how immense these time periods are. We can handle a span of about 250 years, from the birth of your grandparent to the future death in old age of your grandchildren. Ten million years is 40,000 times as long, way beyond our comprehension. As for billions of years, it is a conceptual impossibility.)

Until 1986 nobody thought that any bacteria had circadian clocks. The logic was straightforward. Bacteria grow by cell division. Each cell divides in a process called mitosis. This happens maybe two, three or more times every 24 hours. Ergo, how can a clock that has a period of about 24 hours be of any use to organisms that divide faster? The wrong presumption was that a bacterial cell is equivalent to a sexually reproducing multicellular organism. In truth, a bacterial culture is more like a mass of protoplasm that grows larger and larger and incidentally sub-divides. A mother cell does not die to make daughter cells; she is the daughter cells.

The 'prokaryotic barrier', as Carl Johnson calls it, was broken by Tai-wanese researchers who were investigating how cyanobacteria found in rice paddies remove nitrogen from the air and fix it chemically so that plants can use it for food. Photosynthesis produces oxygen, and oxygen strongly inhibits the nitrogenase enzyme that is vital to capturing nitro-gen from the atmosphere and converting it to ammonia for use in bio-chemical reactions. They found that the unicellular organisms, less than 50 μm in size and with no nucleus, separate out these incompatible processes by photosynthesising during the day and fixing nitrogen at night. 'When those of us who work in biological clocks saw their results, we realised that the algae must have a clock' (Johnson, personal commu-nication). This stimulated a formal search for the biological clock in bac-teria (Kondo *et al.*, 1993).

In hindsight, how simple could it be? Isaac Edery of Rutgers Univer-sity has pointed out, 'Early clocks could have endowed organisms lack-ing spatial barriers with the means to separate incompatible biochemical reactions (e.g., reduction/oxidation or photosynthesis and nitrogen fixa-tion) in time' (Edery, 2000).

The circadian clock in cyanobacteria not only keeps track of circa-dian time in exponentially dividing cells; it also 'gates' the cell's activi-ties. The clock controls the actual timing of cell division. When the gate is open, cell division occurs; but it is forbidden when the gate is closed. In Johnson's phrase, 'cellular events can waltz to a circadian andante even when the cells are rapidly dividing in allegro' (Johnson *et al.*, 1998).

Johnson and his colleagues set up an experiment to test whether these circadian rhythms were in fact adaptive (strictly speaking, it is mechanisms that produce and control behaviour that are selected). They let strains of the cyanobacterium *Synechococcus*, with differing periods, compete between themselves. One strain, the wild type, had a free-running period of about 25 hours; others ran at 22 hours and 30 hours, while a fourth strain was arrhythmic.

In pure cultures, each of the strains grew – or rather they increased through reproduction by cell division – at about the same rate when kept in a 12-hour light/dark cycle. When each of the strains was mixed with another strain and they were grown together in competition, a pattern emerged that depended upon the relationship between the LD cycle and the *Synechococcus* strain's periodicity.

The wild type competed with the 28-hour period strain in a 22-hour cycle (LD 11:11) and a 30-hour cycle (LD 15:15). Wild-type cells grew better during competition in the shorter 22-hour cycle but the longer-period strain defeated the wild type in the 30-hour cycle. When competing against the wild type, the shorter-period strain took over the culture in the 22-hour cycle but lost in the 30-hour cycle. Finally, when the short-period strain was matched with the long-period, the results were clear cut: the strain whose period most closely matched that of the light/dark cycle eliminated the competitor. The other three strains beat the arrhythmic strain on all occasions (Johnson & Golden, 1999).

Although the rhythmic strains could be entrained to the light/dark cycles with which they were challenged, there must be an advantage in having the free-running period close to the environmental light/dark cue. Johnson's experiment was the first rigorous demonstration that there was an adaptive value among competing cyanobacteria in having circadian rhythms with periods close to the current solar cycle. However, showing that present-day cyanobacteria have circadian rhythms does not allow us to say that the fossilised remnants from stromatolites 3.5 billion years old were also rhythmic. But cyanobacteria have not changed much in the past few billion years. Schopf has coined a new word – hypo-bradytely – to describe this ultra-slow evolutionary change. He considers that the cyanobacteria of today probably have the same biochemistry as their very ancient ancestors (Schopf, 1999):

> From early in biologic history to today, the same families, genera, and even species of cyanobacteria have inhabited the same settings, lived in the same kinds of microbial communities, and built the same thinly layered, high-rise stromatolitic condos. This remarkably stable set of relations could have been sustained only if the metabolic lifestyle of cyanobacteria remained unchanged over the aeons.

While the circadian clock in cyanobacteria seems to be of vital importance in maintaining internal temporal order, originally there may have been another factor also selecting for it. Even if the internal metabolism of cyanobacteria has not changed much in a few billion years, the external environment has. Compared with the early earth, it is now positively benign from our point of view. After three billion years of photosynthesis, the atmosphere is oxygen-rich.

There is also a very thin stratospheric ozone (O_3) layer. If it were uniformly layered across the surface of the earth it would be no more than three millimetres thick, but it was non-existent three billion years ago. This gossamer layer of gas absorbs most of the ultraviolet (UV) radiation emitted from the sun. This is important for biological evolution because DNA and proteins are very effective UV absorbers. Damage to DNA is a known cause of malignant mutation, and damage to protein can affect biochemical reactions. We know this well. Four hours on a beach results in about 10 UV-induced errors in the DNA of every skin cell.

Lynn Rothschild and her colleagues at NASA's Ames Research Center in Moffett Field, California, wondered how the early bacteria had survived without even a thin ozone shield. When cells divide, the two DNA strands disentangle and each strand builds a complementary version of itself to produce two double-stranded molecules. The single strand of naked DNA is very susceptible to UV damage.

Her group examined the natural pattern of DNA replication in three present-day microbial mats, two from Yellowstone National Park and one from Baja California in Mexico. At night, they cut off a section of each mat and placed it inside a plastic bag that was transparent to UV radiation. The team then added a small amount of radioactive phosphate, which the bacteria incorporate into any DNA they make. The bags were returned to the mat. The next day, the researchers removed cells from the bags every few hours and measured how much of the phosphate had been used to make DNA (Cockell & Rothschild, 1999; Rothschild & Cockell, 1999).

When the sun rose, the cells went straight to work making DNA. But at noon, the cells shut down the production of DNA for between three and six hours. Production then restarted before sunset. By contrast, the cells photosynthesised throughout the day, suggesting that the afternoon nap applied only to the manufacture of DNA.

These early cyanobacteria had three possible strategies to enable them to contend with the high daytime levels of UV flux and the damage to DNA. They could avoid it altogether; protect UV-sensitive mechanisms, or opt to repair the damage. DNA repair is a vital activity in modern-day cells, but it is difficult to see how the complex multi-enzyme processes of DNA repair could have already evolved 3.5 billion years ago.

Although the easiest avoidance strategy is to live under several metres of water or go underground, neither is particularly helpful to a photosynthesising organism. Furthermore, cyanobacteria are not motile. They could not begin the day on the surface of a water column and then slowly move downwards away from the UV radiation. The only realistic strategy available in practice would have been to time DNA synthesis and cellular division to when potential damage from UV was minimised. This 'escape from light' hypothesis was first put forward by Pittendrigh, but this procedure needs anticipatory capability. If a bacterium had only an acute response to UV, then by the time it responded to the increasing intensity and switched off DNA replication the damage would have been done. It would be more protective to switch off DNA replication in anticipation of the peak in UV intensity.

UV radiation is grouped into three wavelengths, A, B and C, and it is not entirely harmful. Humans rely on it for the production of vitamin D. Lynn Rothschild believes that it also had a positive effect on evolution (Rothschild, 1998):

Natural selection consists of two stages. Generation of heritable variability, in other words, you have to have some sorts of changes that are inherited. And then you have a differential selection of the offspring. Some of them make it and some of them do not and the reasons they make it or do not we assume has something to do with their heredity, not just because they happen to be standing under a tree when it was knocked over by lightning, because that doesn't make any change in the next generation in terms of something that would be of any advantage. It was just bad luck.

With no ozone shield, early life would have been subject to UV-C, the most destructive of the three. The evolution of the circadian clock might have provided the means for striking the fine balance between the variability provided by mutations caused by such UV radiation and its destructive effect on self-copying. Colin Pittendrigh has quoted John van Neumann on the difference between organisation and order. The mathematical genius said, 'Organisation has purpose, order does not' (Pittendrigh, 1993). Pittendrigh took this to mean that the primary purpose of biological organisation is self-perpetuation by self-copying. This is not to say that evolution is purposive. It is not. But there is a point to biological organisation and that point is faithful reproduction. If a timing mechanism enabled the early cyanobacteria to reduce DNA damage from UV radiation there is a strong probability that the mechanism would have been favourably selected and conserved by the organism.

Although the ozone layer protects the present-day earth from the most energetic UV-C radiation, UV-B radiation, which is harmful to cell division, still penetrates. Cells that are naturally exposed to sunlight segregate cell division (mitosis) to the night. *Chlamydomonas reinhardtii*, a nucleated unicellular alga, times cell division, when the DNA is most vulnerable, to late afternoon and evening, when the UV-B flux is attenuated.

If evolution were nice and simple, the story would be something like this ... Once upon a time, early in evolutionary history, the combined selective pressure of UV damage and temporal coordination pushed the development of circadian clock systems in simple organisms. These clocks were so functional and important that they were conserved through evolutionary history and the development of new species and phyla.

It is evident that different phyla have clocks with a similar mechanism. The assiduous reader now should be able to follow Jay Dunlap's technical description (Dunlap, 1999):

> In general, a feedback loop central to these circadian oscillatory systems involves positive and negative elements and is centred on the transcription and translation of clock genes and clock proteins. The positive element in the feedback loop is the transcriptional activation of a clock gene(s) through binding of a heterodimeric transcriptional activator to the clock gene(s) promoter(s). The defining characteristic of this heterodimeric activator is that its two disparate parts are held together via the interaction of protein dimerisation domains called PAS domains. The circadian involvement of such PAS-domain containing DNA-binding positive clock elements was first described in Neurospora (WC1 and WC2) and later in Drosophila (CLK and CYC) and in mice (CLOCK and BMAL).

Although the mechanism is similar in cyanobacteria, the genes in the clockwork are not. For instance, Johnson and colleagues have found what is known as the *kai* cluster of genes in cyanobacteria (Mori

& Johnson, 2001). Early in the day, the *kai* genes are transcribed and translated. Then, KaiA protein turns up the expression of the genes encoding KaiB and KaiC. After a delay, KaiC protein turns off the gene expression of all three. Kai protein levels fall (probably as a function of the cell's protein degradation systems); thus as KaiC protein levels fall, the genes all get expressed again and the 24-hour cycle repeats. Takao Kondo of Nagoya University says, 'this is a feedback-loop model for the circadian oscillator' of cyanobacteria (Kondo, 1998).

Deleting each of these genes separately (*kaiA, kaiB, kaiC*) affects the rhythmicity of the organism but not its viability, so it looks as though they are key components of the clockwork. Kondo has found strong candidates for homologues to the *kaiC* gene in some Archaebacteria. However, there is no direct evidence yet that the Archaebacteria have an endogenous daily clock.

Even when the clock genes are similar, as in insects and mammals, they do not have identical functions, and the feedback loops are put together differently. On the other hand, there has obviously been conservation of the clock mechanism, for example within the vertebrates. Menaker believes (Menaker & Tosini, 1996) that

> *In both birds and lampreys the photoreceptive pineal gland contains circadian oscillators that regulate the rhythmic production of the hormone melatonin which in turn may be responsible for the circadian aspect of locomotor behaviour. It stretches credulity to believe that these detailed similarities in biochemistry (melatonin), morphology and physiology (the pineal) and behaviour (locomotion) have arisen by convergent evolution.*

Classic examples of convergent evolution are the body shape of dolphins, sharks and penguins, or the red tube-shaped flowers of plants pollinated by hummingbirds. Different organisms can look alike but have very different ancestries. A streamlined shape is essential for fast swimming and so a dolphin (mammal), shark (fish) and penguin (bird) have converged to a similar body shape that is required for the ecological niche each occupies. Likewise, plants that are pollinated by hummingbirds look alike even though they are derived from different ancestral lines.

Menaker is saying that the circadian clock in vertebrates is not a case of convergent evolution but pre-dates their emergence 600 million or so years ago. The differences in the detailed clock mechanism between, say, a bird and a lamprey are not due to its being reinvented time and time again. What differences exist are more in the nature of adjustments than fundamentally different arrangements.

Insects share a common system with vertebrates. Although the clock mechanism in *Drosophila* is not the same as in a mouse, some of the genes are homologues and have a common ancestry. So there was one ancestral clock present before the invertebrate/vertebrate split. But it looks as though the clock evolved separately in cyanobacteria, fungi and animals. There is an interesting suggestion that the circadian mechanism came to eukaryotes pre-packaged in a proverbial single bound. Some 30 years ago Lynn Margulis proposed what she called the serial endosymbiosis theory as a way of explaining the peculiar genetic independence of chloroplasts and mitochondria. She suggested (Margulis, 1998) that

> *molecular biological, genetic and high-powered microscopic studies all tended to confirm the once-radical nineteenth century idea that the cells of plants and of animal bodies (as well as fungi and all other organisms composed of cells with nuclei) originated through a specific sequence of mergers of different types of bacteria.*

According to Margulis and her colleagues, the photosynthesising chloroplasts were free-living bacteria that had taken up residence inside the cells of many yeasts, protists and even plants and animals. Perhaps the clock mechanism of an ancestral cyanobacteria was passed by endosymbiosis into a host that later became eukaryotic plants.

Nice as the idea is, the evidence so far does not support it. *Arabidopsis* is the model plant for circadian studies and no homologue of the *kai* genes in the cyanobacteria has yet been found.

In summary, either oscillatory function is highly conserved and evolutionarily ancient, or it is easily acquired and has evolved multiple times. In fact, both scenarios seem to be true. There is evidence of structural and functional homology of the oscillator between model systems (such as mouse and *Drosophila*), as well as an emerging picture of how circadian rhythm systems originated by the appropriation of a variety of cellular mechanisms for the task of structuring a biological clock.

Although the basic mechanism shown in Figure 10.1 seems simple

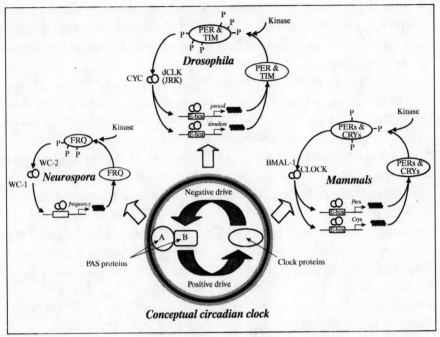

Figure 10.1. The circadian clock in mammals, *Drosophila* and *Neurospora* shares the fundamental property of a negative feedback loop. The conceptual circadian clock consists of two proteins, A and B, that form a protein complex, connected to each other via their PAS domains. This protein complex provides the transcriptional drive (positive drive) that leads to the production of a series of clock proteins. These proteins then provide the inhibition (negative drive) that either shuts off or reduces the transcriptional drive by the PAS proteins A and B. The stability of the clock proteins, and their ability to enter the nucleus and inhibit the PAS protein complex, determine the period of the oscillation. Protein stability is often regulated by its level of phosphorylation (P) by protein kinases. After phosphorylation proteins are usually targeted for destruction. These basic elements are seen in the mammalian, *Drosophila* and *Neurospora* clock mechanisms. In *Neurospora*, the WC-1:WC-2 PAS proteins provide the positive drive on transcription, while FRQ protein provides the negative drive. In *Drosophila*, the positive drive is provided by the CYC:CLK (CLK is also known as JRK) protein complex, and the negative drive by the PER:TIM complex. In mammals, the positive drive comes from the BMAL1:CLOCK complex and the negative drive by interactions between PER and CRY proteins.

enough, the devil is in the details. The very complexity of the models in Chapter 7, with their cycles and epicycles of positive and negative feedback loops, is now beginning to look as ungainly as Ptolemaic versions of the solar system.

There are some very serious questions about the orthodox explanation of the circadian mechanism. At its heart is delayed negative feedback: the direct or indirect inhibition of a process by its products. Equally important is the positive drive to the system, against which negative feedback can exert an effect (Figure 10.1). Despite all appearances, the case for this is not proven. Nor are we certain even that the clock genes themselves are critical to the circadian system. We have seen in *Drosophila* that one protein, PER, expressed by the *per* gene, is a vital component of the feedback loop. Therefore, it would follow that *per* is important in circadian rhythmicity. But the PER protein it expresses is found in many different tissues and in some of them at least it is not found in the nucleus, and neither is it rhythmic. In other words, *per* expresses PER and in some instances it is rhythmic and others not. Patricia Lakin-Thomas is one of those who has raised the question: 'Clock genes must have other functions that are not explained by the model and this raises the possibility that these proteins might have a primary function that is as yet unknown and that their effects on circadian clocks might be secondary' (Lakin-Thomas, 2000).

In the simple circadian feedback loop in *Neurospora*, the positive element in the loop comes from the two PAS-protein transcription factors, WC-1 and WC-2. They form a heterodimer that activates transcription of the *frq* clock gene. This clock gene *frq* is rhythmically transcribed in the nucleus and translated into its protein FRQ in the cytoplasm. The protein is then transported into the nucleus, where it inhibits transcriptional activation by the heterodimer WC-1:WC-2. Light is another positive element, as it activates transcription of *frq* by interacting with WC-1 and WC-2 (Figure 10.2).

The key fact is that this is a closed loop, composed of positive elements driving transcription and translation of a gene that is then subjected to negative feedback by the products of that gene's transcription and translation. There is no doubting the importance of *frq*. Without it, the circadian rhythmicity of *Neurospora* loses key characteristics. It is no longer self-sustainable in constant conditions; it is not entrainable by light. Yet there are strains created by laboratory mutagenesis that do not

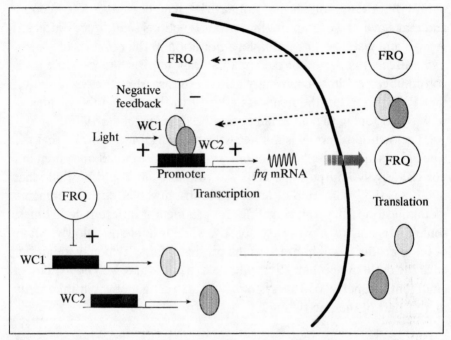

Figure 10.2. The *Neurospora* feedback loop. WC-1 and WC-2 drive *frq* transcription in the presence of light. FRQ provides the negative feedback inhibition of its own transcription. WC-1 and WC-2 are not expressed rhythmically. FRQ also promotes the production of WC-1. Not all of the proteins of the *Neurospora* transcriptional/translational feedback loop have been shown.

have the *frq* gene and are still rhythmic, although with a greatly altered periodicity (Merrow *et al.*, 1999).

Likewise, four photoreceptor genes in *Arabidopsis* – *phyA*, *phyB*, *cry1* and *cry2* – can be 'knocked out' in a quadruple mutant, yet the plant remains entrainable by light (Somers *et al.*, 1998). So the search is on for more light inputs, and several candidates have been found. It is a case of genes piling on genes and the model becoming more and more complex.

A theory of circadian timing has to reconcile a self-sustaining rhythm generated by a group of clock genes with the uncomfortable fact that removing one or more of these genes does not necessarily destroy this very same rhythm. Because it has been assumed that clocks must be built from special proteins dedicated to time-keeping, the mutation-screening work looked at those features that affect the clock, such as the period. But these proteins may not have been dedicated to time-keeping and may have had even more important functions that directly affected growth and viability. For example, a mutation in the *cel* gene of *Neurospora* affects the period of the circadian rhythm. But this *cel* gene is involved in critical long-chain fatty acid synthesis. Also, the *tau* mutation that Ralph found in the hamsters alters growth and metabolic rates as well as shortening the circadian period (Oklejewicz *et al.*, 1997).

The orthodox model is not necessarily wrong, but it may be that the clock genes are an epiphenomenon, involved in the circadian system but not exclusively and perhaps not even primarily. It could be that what matters is the interaction of feedback loops, however derived. Imagine metabolic loops all producing different products at different cycle times; some loops may turn over very quickly, while others quite slowly. Many of these products will interact, as the product of one loop may well influence the cycle of another. What we see as a clock may well be akin to an emergent property of the system and all the genetic paraphernalia merely part of the fine tuning.

11

SLEEP AND PERFORMANCE

*Whatever physiological variables we measure, we usually find that there is
a maximum value at one time of day and a minimum value at another.*

JÜRGEN ASCHOFF (ASCHOFF, 1965)

Second-year physiology students at Imperial College, London, are introduced to the importance of circadian rhythms when they monitor their
rectal temperature regularly over a 24-hour period (not as uncomfortable as it sounds). They find their body temperature fluctuates rhythmically with a peak-to-trough change of about 1°C. The low point is around
4.00–6.00 a.m. and the high-water mark is about 12–14 hours later.

The sleepy feeling in the early afternoon that we know as the post-
lunch dip is also due to our circadian biology, although a large lunch and
a few drinks compound the feeling. Many of the students also introduce
themselves to the difficulties of sleeping after all-night partying. The
smarter ones soon learn that sleep, naps, alertness, mood, body temperature and circadian rhythms are intimately connected.

The first well-controlled attempt to study the daily changes in
human efficiency and alertness began in the 1930s with the work of
Nathaniel Kleitman. He showed that the performance of a wide range of
factors such as multiplication, hand steadiness, and card dealing and
sorting tended to rise in the morning and then fall after reaching a peak
around the middle of the day (Kleitman, 1963).

Kleitman is universally recognised as the father of sleep research.

177

He died in 1999 aged 104. When he was well into his nineties he was asked about his most famous book, *Sleep and Wakefulness*. Kleitman said, 'when I came to write my book *Sleep and Wakefulness*, I put the word Sleep first because the library and catalogues did not have any heading for Wakefulness or Awake.' The monumental work lists over 4,000 references and in the days before the PC Kleitman typed out every file card himself. It took him longer than the writing of the book (Rossi, 1992).

In an epic and highly publicised experiment, he and an assistant, Bruce Richardson, lived 400 metres underground in Mammoth Cave, Kentucky, for over a month in 1938 to see how they would get on. Though Kleitman said he could not sleep properly, Richardson adapted. In the cave, Kleitman measured body temperature and discovered that there was a slight but regular fluctuation throughout the day, and that peak efficiency occurred when body temperature was highest (Kleitman, 1963).

There is a view that without quite realising it Kleitman had done enough to show there was an endogenous clock. As William Dement at Stanford University notes, 'Had both individuals demonstrated an apparent fixity of the 24-hour rhythm, Kleitman might have turned in the direction of postulating an internal clock' (Dement & Vaughan, 1999).

By the early 1960s, the obvious discomforts caused by night working and its potential health impairment were troubling researchers. They began contemplating a world in which round-the-clock operations were becoming far more common in power plants, utilities, data processing, aviation, news media, telecommunications, the military and manned space flight. Jürgen Aschoff, a professor at the Max Planck Institute for Behavioural Physiology, was working on some of these problems. Living in caves was somewhat extreme and hardly an ideal environment for careful scientific studies. He had built a prototype 'time-free environment' in a former basement shelter under a Munich hospital and chose himself as the first subject. A stomach complaint shortened the stay, but he was in the cellar long enough to show that a human subject in isolation from time cues showed endogenous free-running rhythms, just like a rat (Aschoff, 1960).

Research students keen to write up a thesis and be paid in the process have long been experimental fodder in extended sleep experiments, but even for them the makeshift cellar in Munich was barely adequate. A more serviceable version of Mammoth Cave was needed.

Figure 11.1. An original actogram from a human subject in a study by Aschoff and Wever. For each day, sleep (the solid line) and wake and activity (the dotted line) are indicated. For the first nine days, the male subject was in the normal environment of daily living with alternating light and darkness, warmth and cold, noise and quiet. These physical and social cues change rhythmically and

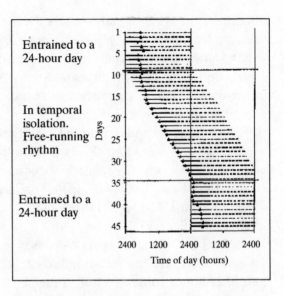

entrain the subject's endogenous circadian oscillation to a periodicity of exactly 24 hours. This is clearly shown in the first period of the actogram, with the subject going to bed and getting up at about the same time each day. The solid triangles record the low point of the subject's core body temperature during a 24-hour period, and in the first nine days it is near the end of the sleep period. After the first nine days, all time cues were removed from the subject's environment. Temperature, humidity and noise levels were all kept constant. He controlled the light himself, prepared his own meals when he was hungry, and went to sleep whenever he chose. But just like the rat held in constant darkness, there were no external cues that linked the subject to the daily rotation of the earth. The circadian oscillation in body temperature and sleep/wake cycle continued but the oscillation was no longer exactly 24 hours. In an isolated environment, the subject was free-running. Under these free-running conditions the body temperature rhythm changed its phase relationship to the sleep/wakefulness rhythm. Instead of reaching its low point towards the end of sleep, the low point occurred at the beginning of the sleep period. The rhythms became uncoupled, showing internal desynchronisation. In the the lower section of Figure 11.1, the individual was exposed to a 24-hour light/dark cycle, with lights off at midnight and turned on around 8.00 a.m. Meals were served at specific times each day. The subject entrained and bedtime become stable at around midnight, and the low point of body temperature moved back towards the end of the sleep period (Aschoff *et al.*, 1967).

Pittendrigh and Aschoff had developed a friendship that was to last until Pittendrigh's death, and on his advice NASA funded the building of an underground facility at Andechs in Bavaria for research on human circadian rhythms. For the next 20 years, Aschoff's department was a leading centre and he is rightly regarded as one of the group who established chronobiology as a serious discipline. With his co-worker Rutger Wever, Aschoff designed an underground facility wrapped in copper wire so that the subjects would be shielded from possible external time clues arising from daily electromagnetic variations. Male staff shaved at odd hours so that the subjects could gather no idea of the time from the state of a man's beard. The results of a typical study are shown in the stylised actogram in Figure 11.1.

Aschoff and his co-workers did this type of experiment many times and concluded that the free-running circadian rhythm in humans was close to 25 hours, somewhat longer than in other species. But something about the experimental set up troubled Charles Czeisler, then a young researcher, when he first visited the Max Planck Institute facility in 1976. Now a professor of medicine at Harvard, Czeisler recalled (Cromie, 1999):

These experimenters allowed their subjects to switch on lights when they were awake and turn them off when they wanted to sleep. They didn't think this would have any effect, but switching on electric lights resets the biological clock. It's the same as resetting your watch.

Unlike mice or rats, humans cannot be kept in complete darkness for very long so it was impossible to replicate the classic studies on animals that were kept in light/dark cycles and then allowed to free-run in the dark. Experimental protocols designed to determine the free-running rhythm have to allow for the fact that humans need some light exposure. But the Andechs experimenters clung to the view that light was not important as a *Zeitgeber* in humans. They were wrong-footed because of an accident in an early experiment. Aschoff and his colleagues marked the change from the light to dark cycle with the sounding of a gong to remind the subjects to carry out various procedures such as urine collection or performance tests. On one occasion, however, the auditory system failed and surprisingly the subjects went into a free-running rhythm, even though they were still subject to a light/dark cycle. The subjects later explained that the auditory signal was regarded as a form of 'social contact' with the investigators. When this 'social entrainment' was confirmed in repeated experiments, Aschoff and his colleagues decided that humans reset their timing mainly by social cues and that we were the exception to the rule that light is the principal synchroniser (Aschoff *et al.*, 1971).

It took Czeisler something over 20 years and many experiments to overturn successfully the view that the human circadian system is insensitive to light and that the human clock runs at 25 hours. He and his colleagues showed that notwithstanding the importance of social cues in the entrainment of the human clock, even the low light levels in offices and homes did in fact have an effect. Czeisler did it by effectively creating a 28-hour day for his subjects by making them go to bed four hours later each day. This strategy disconnected the biological pacemaker from clock time, as he explained (Cromie, 1999):

The 28-hour cycle distributed light exposure, sleep and wakefulness, work and play evenly around the biological clock. The men and women did not get light exposure at the same time each clock day. Instead, they experienced a six-day week in which light and dark occurred at different times each day.

Over a month, the daily rhythms of hormones and body temperatures in 24 healthy young and old men and women rose and fell on an average cycle of 24 hours and 11 minutes, slightly longer than 24 hours, but significantly shorter than past estimates of 25 hours. Czeisler had been right all along: the low-intensity light left on at night had interacted with the human circadian clock and so given a false reading of the endogenous rhythm (Czeisler *et al.*, 1999).

A few minutes may seem neither here nor there, but the human circadian period is a very important factor in the planning of the proposed manned trip to Mars. It is part of a wider problem affecting astronauts, who are subjected in space to daylengths that differ from the familiar 24-hour cycle. On shuttle missions astronauts on the shuttle's flight deck experience 45 minutes of light, succeeded by 45 minutes of dark. The shift schedules on the shuttle typically operate on 23.5-hour days and many astronauts experience sleep difficulties, averaging only about six hours of sleep a day, in contrast to the seven or eight hours they get on the ground (Dijk *et al.*, 2001). This can lead to an increased risk of accidents due to fatigue and sleepiness. After 90 days in space, the endogenous clock of astronauts on the Mir mission and the International Space Shuttle began to wane, and sleep quality and quantity deteriorated.

Although the Martian day at 24.65 hours is only 4 per cent or so longer than the normal 24.14-hour human circadian day, this makes it difficult for the human clock to entrain. Kenneth Wright and his colleagues evaluated how people's internal clocks were affected by exposure to 23.5-, 24- and 24.6-hour days in a NASA-funded earth-based project. Since spacecraft cabins are dimly lit, study participants were exposed to low-level daytime lighting equivalent to candlelight. All groups were placed on a fixed work/rest schedule. The phases of the subjects' rhythm of pineal melatonin were evaluated to determine how they adapted to the various daylengths. In a normal day/night sleep cycle, melatonin levels rise about two hours before an expected sleep period, are high during sleep and are of course low during the day. Participants on the 24-hour light/dark schedule maintained the appropriate phasing of the melatonin rhythm for sleep. Those exposed to 23.5-hour or 24.6-hour cycles showed free-running rhythms. Wright said, 'The groups experiencing the shorter or longer days did not adapt' (*Spaceflight Now*, 2002).

Those individuals exposed to a 23.5-hour or 24.6-hour light/dark cycle showed an abnormal phasing of their melatonin rhythm. Levels were often high when the subjects were awake and low when they were trying to rest. This factor made it difficult to sleep at the scheduled time. 'This problem with the melatonin cycle occurs during jet lag and in people working on night shifts,' according to Wright. 'In effect, astronauts on shortened days are experiencing jet lag in space' (*Spaceflight Now*, 2002). Enter rocket-lag!

On Mars, every 41st day the astronauts could be 12 hours out of phase with the Martian day, resulting in impaired performance of daytime alertness – not something you would want to happen on the surface of Mars. The problem is compounded because on Mars the daylight is primarily yellowish-brown. On earth it is blue-green. This could make a difference because the photoreceptors entraining the circadian clock are more sensitive to the 'bluer' wavelengths of light than the redder wavelengths that dominate the Martian sky.

We recognise sleep when we see it in others or even in other species.

During sleep, humans and other animals show specific postures, display minimal movement and have reduced responsiveness to outside stimuli such as noise (Hendricks, 2003). Of course this condition is reversible, and this distinguishes sleep from other states like coma, anaesthesia or even death. Sleep is studied in the laboratory by monitoring the pattern of an individual's brain waves or their electroencephalogram (EEG). When awake the EEG shows a high-frequency low-amplitude oscillation in electrical activity. But during the onset of sleep and the descent into slow-wave sleep, the EEG is characterised by an increase in the amplitude and a decrease in the frequency of the oscillations.

Sleep moves progressively through four recognised stages into deep slow-wave sleep (stage 4). However, after approximately 70 minutes the sleep state reverses, and stage 4 sleep moves rapidly up to stage 3, then 2, then 1. Yet at the beginning of the night an individual does not normally wake at this point but passes into another state of sleep. In this state, the EEG is very similar to that in the awake brain, with high-frequency and low-amplitude electrical activity, but the individual is not awake. The eyelids are shut, but the eyes start to move rapidly. Heart rate and blood pressure increase, and the body is held almost completely still. This is called rapid eye movement sleep, or REM sleep. After some minutes of REM sleep, there is a switch back to non-rapid eye movement, or NREM, sleep. The individual will pass down through stages 1–4 and then back up to REM sleep once again. This cycling of NREM and REM sleep lasts approximately 90 minutes, and in an average night we may experience five of these NREM/REM sleep cycles (Dement & Vaughan, 1999)

These cycles are not identical. The first part of the night is characterised by the occurrence of deep slow-wave NREM sleep (stages 3 and 4), and in the second part of the night we undergo more frequent and longer episodes of REM sleep. We usually wake naturally from REM sleep. Dreams are most frequently remembered when an individual is woken from REM sleep, but can also occur during the different stages of NREM sleep. For centuries, sleep has been regarded as a simple suspension of activity; today we appreciate that it is a complex and highly organised series of physiological states, but to what end?

As yet nobody knows why we sleep, but there is no dispute that sleep is essential to well-being. For reasons that are not at all clear, rats deprived of sleep die between 14 and 40 days of deprivation. There have been instances of human sufferers of familial fatal insomnia, who are unable to sleep and who also die. FFI is a very rare disease occurring in perhaps 20 or so families world-wide. One type is associated with a defect in a prion protein that when triggered is invariably fatal after about 6–30 months without sleep. The pathological processes include degeneration of the thalamus and other brain areas, overactivity of the sympathetic nervous system, hypertension, fever, tremors, stupor, weight loss, and disruption of the body's endocrine systems (Scaravilli *et al.*, 2000).

While the effects of sleep deprivation may be known, there is no consensus as to how much sleep we actually need. A much-touted American study of over one million adults found that subjects sleeping less than four hours a night were associated with an increased death rate (Kripke *et al.*, 2002). Those who slept seven hours a night had the best survival rates, taking into consideration such variables as age, diet, exercise, previous health problems and known risk factors such as smoking. According to the report, the group sleeping eight hours or more a night were 12 per cent more likely to die over the course of the study than were those sleeping seven hours a night. Quite what to make of this is not clear, as it is probable that even in this controlled study at least some of those sleeping eight hours or more may well have had an ongoing disease. One result could be that the increased sleep is a reflection of co-morbidity – when you are not well you sleep longer anyway.

Although the finding of a detrimental effect of more than eight hours' sleep is controversial, it does seem that on average adults need about six to seven hours sleep a night to be well rested and fully functional. Reducing nocturnal sleep by as little as one hour increases physiological sleepiness the next day. But we can manage on less, and the rule-of-thumb advice is that if you are not falling asleep in the middle of the day then you are probably getting enough sleep. However, the effects of successive nights of reduced sleep accumulate. If you need seven hours of sleep a night and through the working week you only average six hours, then by Friday you are four hours 'in debt'. As a consequence, by Friday you will be sleepier, and functioning more slowly and less reliably, than on Monday. Your mood may also become more irritable and negative. A sleep camel is a person who makes a habit of getting little sleep during the week and tries to make up for it by napping and sleeping in over the weekend. It is a hopeless task, according to David Dinges at the University of Pennylvannia, who makes the point that two, three or possibly even four or more nights of oversleeping may be required to reduce the sleep debt to zero and recover the waking alertness degraded by that debt (Buysse *et al.*, 2003).

The trouble comes when an increasing sleep debt conflicts with the circadian sleep/wake rhythm and, tired as you are, sleep does not come easily. A wild Friday night out may end at five in the morning with you tucked up in bed, but the chances are that after four or five hours of poor-quality sleep you will be awake and unable to get back to sleep. There is a conflict between the sleep drive that wants you to recoup the sleep debt and the circadian rhythm that is saying, metaphorically of course, 'hey, get up, it is daytime and you should be out and about.'

Building on the work of Alex Borbely (Borbely, 1982), William Dement, a one-time student of Kleitman, has suggested that sleep is generated by two broadly opposing mechanisms. There is a circadian-driven rhythm for wakefulness that promotes wakefulness throughout the day, and then decreases the drive for wakefulness during the sleep phase (Figure 11.2A). This is opposed by what has been described as a 'homeostatic drive for sleep'. The homeostatic drive describes an intu-itive process whereby the drive for sleep increases the longer an individ-ual has been awake (Figure 11.2B). These two processes interact to consolidate sleep (Figure 11.2C). In Dement's words, 'the push and pull of these opposing processes allows us to stay up all day and sleep all night' (Dement & Vaughan, 1999). Homeostasis maintains the duration and intensity of sleep, while the circadian rhythm determines the timing of the propensity to sleep.

Dement suggests that in effect we have a metaphorical accountant inside us meticulously keeping a record of how much sleep we have had in the preceding period and how far short of the desirable norm we are. When our sleep debt is zero we wake, but on awakening we immediately begin accumulating a sleep debt that pushes us towards sleep. Opposed to this sleep tendency is what he calls clock-dependent alerting. This is the circadian rhythm that pushes us to wakefulness during the day but is inactive at night.

In the first few hours after wakening, our sleep debt is very low and so is our sleep drive (Figure 11.2). At this time the circadian drive for wakefulness does not have to be very strong to keep us alert. As the day progresses, the result of the tension between the drives is a biphasic pat-tern of daytime alertness in humans, with a 'midday dip' in alertness

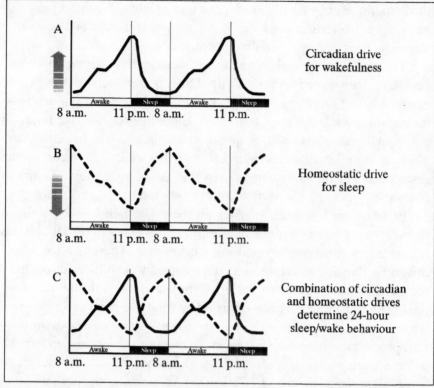

Figure 11.2. The opponent-process model for sleep. Sleep is generated by two broadly opposing mechanisms. (A) The circadian drive for wakefulness. During the day the drive for wakefulness increases progressively. The open bar represents wakefulness and the black bar sleep in each of the figures. (B) The homeostatic drive for sleep. The longer an individual has been awake the greater the homeostatic drive for sleep. (C) The two opposing processes for wakefulness and sleep combined.

occurring in the mid-afternoon. Whereas many societies used to allow a daytime nap or siesta, our modern, globalised society increasingly forces us to use other behavioural or pharmacological measures to get us through this down period (Broughton, 1998). A jog, a blast of fresh air or a cup of caffé latte with a double shot of caffeine is our siesta substitute. By late afternoon the circadian wakefulness drive is in full swing. This keeps us going against the rising sleep debt and at 10.00 p.m. we are surprisingly alert, even more so than earlier in the day, despite having been awake longer.

In the absence of the circadian component, after an SCN lesion for example, sleep will still occur, but it becomes highly fragmented and is expressed as a continuous series of relatively short sleep episodes that are promoted by the homeostatic drive alone (Cohen & Albers, 1991).

Derk-Jan Dijk, a sleep researcher now at the University of Surrey, explains the functional significance of the process: 'by providing a drive for wakefulness which becomes progressively stronger in the course of the waking day, the circadian pacemaker counteracts the progressive drive for sleep associated with sustained wakefulness.' Dijk has suggested that the SCN may not only promote wakefulness during the day, but may also actively promote sleep at night, thereby explaining nocturnal sleep maintenance. After all, most of us will have satisfied the sleep deficit by about 3.00 or 4.00 a.m., yet we sleep through to 7.00 a.m. or later. 'The progressive increase in the circadian drive for sleep in the course of the nocturnal sleep episode counteracts the dissipation of sleep drive associated with consolidated sleep' (Dijk, 1996).

The opponent-process model put forward by Dement and Dale Edgar is a development of an earlier model that suggests that although alertness was once thought to be a cognitive choice, the need for sleep is in fact a biological drive. Dissecting out the relative contributions of the sleep drive and circadian wakefulness drive involved subjecting human subjects to artificial sleep/wake periods. Human subjects were put in bed and allowed to sleep for only one hour out of every three, for several days around the clock. Despite severe sleep deprivation accumulating over several days, subjects were still unable to sleep during the periods in the afternoon and evening that coincided with the peak of the circadian cycle when they would 'normally' be awake (Dijk & Czeisler, 1994).

Everybody has an individual profile or chronotype that describes their rhythmic behaviour over 24 hours. The profile will vary from individual to individual. Some people are larks and are most active in the morning, whereas the owls prefer later in the day. Larks tend to be older; college students and twenty-somethings are notorious owls. Larks are most aware around noon, work best in the late morning and are chatty, friendly and pleasant from about 9.00 a.m to around 4.00 p.m. Owls, on the other hand, do not really get going until the afternoon, are at their most pleasant (if that is not an oxymoronic term for college students) later in the day and are at their most alert about 6.00 p.m.

Perhaps 1 in 10 of us are what is known as extreme larks and another

1 in 10 are extreme owls. In their book on the effects of the body clock on human health, Michael Smolensky and Lynne Lamberg describe as hummingbirds the 80 per cent who are neither extreme larks nor extreme owls but are somewhere in between (Smolensky & Lamberg, 2000). Early-rising larks have usually adopted a smug moral superiority based on Benjamin's Franklin's maxim 'Early to bed and early to rise makes a man healthy, wealthy, and wise.' But there is no basis for the claim. Catharine Gale and Christopher Martyn of Southampton University followed up a 1973 survey that had included data on sleeping habits. More than 20 years later they found no evidence among the survivors that 'following Franklin's advice about going to bed and getting up early was associated with any health, socioeconomic, or cognitive advantage. If anything, owls were wealthier than larks, though there was no difference in their health or wisdom.' In a wry aside they offer the thought (Gale & Martyn, 1998):

It seems that owls need not worry that their way of life carries adverse consequences. However, those who cite Franklin's maxim to encourage their children to go to bed early may wish to consider whether their practice is entirely ethical.

This evening/morningness is less a matter of choice than it is of genetics. Being bright-eyed and bushy-tailed first thing in the morning and raring to go is not just a case of how much sleep someone has had, nor is it a reflection of willpower. The genes may largely determine it. Louis Ptácek and colleagues at the University of Utah studied 29 people in three families with familial advanced sleep-phase syndrome (FASPS). One family included a grandmother, daughter and grandchild all with the same sleep disturbance. Those with the disorder have a shorter than normal wake/sleep cycle. Regardless of work schedules or social pressures, they cannot stay up much later than 7.30 p.m. and they tend to wake up around 3.30 a.m.

The disorder shifts the normal wake and sleep pattern forward by three to four hours. By studying the family relationships, Ptácek found that the disorder is inherited and in some patients, but not all, the problem is caused by a mutation in the human *Per2* gene on chromosome 2. A glycine amino acid appears where a serine amino acid should be. This single alteration occurred in the portion of the h*Per2* protein that governed binding to an enzyme called casein kinase one-epsilon (CK1ε).

hPer2 is homologous to *Drosophila* and hamster genes that when mutated are known to speed up the circadian rhythm (Toh *et al.*, 2001). The University of Utah lab showed that the mutation disrupted phosphorylation of the *hPer2* protein by CK1ε. Ptácek offers the possibility that the disruption in phosphorylation of the *hPer2* protein may prevent the protein from acting as a sort of rheostat that helps adjust the length of the circadian rhythm according to the amount of phosphorylation. 'Circadian rhythm is probably governed by a balance of phosphorylation of different proteins,' said Ptácek; 'in a normal, healthy twenty-four-hour clock, many proteins are being phosphorylated by casein kinase one epsilon, and it is the balance of one protein versus another that produces the normal rhythm' (Ptácek, 2001). However, not all people with FASPS have this mutated gene and it is possible that other genes may be involved, particularly those that are associated in the melatonin rhythm.

Apart from its importance in helping to understand the relationship between the circadian clock and the sleep process, the work on FASPS is the first time that scientists have uncovered a genetic mutation leading to a change in a complex human behaviour such as sleep. It parallels Benzer and Konopka's finding that a single mutation in *Drosophila* had a profound effect on the insect's rhythmic behaviour. It could also have profound societal consequences for us. Ptácek notes (Ptacek, 2001):

> *We can think of major disasters, such as Three Mile Island, where a conflict of sleep patterns and work schedules may have contributed to the accident. But we can also think of everyday problems such as traffic accidents and insomnia. Even school schedules can be difficult for children who can't get up early in the morning.*

Commenting on the discovery of the first FASPS family by Chris Jones and colleagues at the University of Utah, Cliff Singer and Ali Lewy wryly noted, 'It seems that our parents – through their DNA – continue to influence our bedtimes' (Singer and Lewy, 1999).

About 1 in 14 adolescents cannot get out of bed in the morning. These late risers are neither lazy nor fractious. He or she, though more usually he, may well have delayed sleep-phase syndrome (DSPS). Insomnia prevents him or her from getting to sleep at a reasonable time. Sufferers complain of excessive sleepiness and are unable to fall asleep or to wake up at the desired time. William Dement, who has studied the disorder, says (Dement & Vaughan, 1999):

> *Their efforts to advance the timing of sleep onset such as going to bed early, having a friend or family member get them up in the morning, trying relaxation techniques or using sleeping pills is not permanently successful. They often describe sleeping pills in normal doses as having little or no effect in helping them fall asleep.*

As yet, there is no detailed understanding as to the cause of DSPS. Like FASPS, it is a persistent fault in the phasing of the circadian rhythm, in the same way that a pulse of light administered at specific phases in the circadian cycle can advance or retard the rhythm. Sufferers from FASPS or DSPS have a defect that expresses itself in a similar manner to that advancement or retardation. In some ways it is similar to jet-lag, in that the body's time is out of synchronisation with local time, but whereas jet-lag is transient and wears off in days, sufferers from FASPS and DSPS are stuck with it. There has been some success with putting FASPS sufferers in front of a bank of very bright lights in the evening for about two hours, and early in the morning for those with DSPS. It is a difficult schedule to maintain and the results are not always satisfactory. However, unlike the genetically hard-wired FASPS, young people seem to grow out of DSPS as they leave their teens (Okawa *et al.*, 1998).

While parents tear their hair out getting their teenage children up and off to school, teachers have long known about the intimate connection between sleepiness and their students' alertness over the course of the day. The study of alertness and performance rhythms dates back to the early days of experimental and educational psychology, when there was a keen interest in determining the optimal time of day for teaching an academic subject. Nathaniel Kleitman pointed this out some 50 years ago: 'the individual's capacity for doing mental or physical work is not the same throughout the waking period', and he went on to emphasise that this 'has been known for a long time' (Kleitman, 1963).

Rhythmic behaviour consists of an endogenous component that is determined by the period of the rhythmicity within the internal clock and an exogenous component driven by lifestyle and environment. Our ability to perform mental and physical tasks changes during the course of the day depending not only on what we are doing at any given time but also on what we have been doing in general in the hours and days before. These outputs, ranging in humans over everything from hand grip strength to the ability to do addition, vary during the day to such an extent that Patrick O'Connor, an exercise science professor at the University of Georgia, concludes, 'Some physical and cognitive performance measures can vary by as much as 15 percent over the course of a day' (personal communication). Kleitman was convinced that the performance factor rhythm not only paralleled that of body temperature but was in fact caused by it. In other words, a single circadian rhythm that determined body temperature in turn determined the performance rhythm.

Others were reluctant to go that far. The major difficulty in studying the effect of time of day on alertness and performance rhythms is untangling to what extent changes are a reflection of tiredness or sleep debt and to what extent are they rhythmically driven by the circadian timing system. Dale Edgar was able to separate the circadian and sleep drives when he lesioned the SCN in squirrel monkeys and showed that although the animals became arrhythmic they slept in total more than before (Edgar *et al.*, 1993). Instead of sleeping for about eight or nine hours in one consolidated bout, the monkeys kept falling asleep for short periods. Overall, they slept for about 12 hours and this was compelling evidence that the circadian pacemaker generates the alerting signal that keeps the squirrel monkeys awake for 16 hours, and that this cannot be done in the absence of the SCN (Figure 11.2).

But we cannot just chop out the SCN in humans, however humanely. Occasionally, a person has a lesion in the SCN area, either by accidental damage or in the excision of a tumour. In those cases, the subjects do become broadly arrhythmic, but relying on this opportunistic method of investigation is hardly the basis for an experimental protocol (Cohen & Albers, 1991).

The problem is compounded by the wide range of lifestyle and environmental factors that can influence the outcome. For instance, exercise obviously raises body temperature while sleep reduces it. So any measurement of the circadian rhythm of body temperature would have to allow for these masking effects. Controlled experiments with humans to separate out the homeostatic sleep drive from circadian alertness are difficult because considerable care has to be taken to remove the masking effects of sleep itself (by allowing, say, nine hours' sleep and then keeping the subjects awake for 19 hours in a 28-hour day), variations in ambient light and temperature (by using constant levels of low illumination and of environmental temperature), activity levels (by keeping the subjects recumbent) and large meals (replaced with small snacks).

Sorting out what is due to time of day from simple tiredness is hard enough for physiological rhythms such as temperature. It is much harder for the corresponding changes in mood and performance efficiency. The core body temperature is always the same irrespective of how it is measured. This is not so for performance rhythms because the subjects under test 'learn' how to do the test with practice.

Yet we all know instinctively that we are better at doing some things at certain times of the day and other things at other times, and many studies in the lab have shown this. As Simon Folkard at Swansea University put it, 'Perhaps the main conclusion to be drawn from studies on the effects of time of day on performance is that the best time to perform a particular task depends on the nature of that task' (Folkard, 1983).

That was in 1983. At the time, the general view was that simple processing tasks tracked body temperature, so that the ability to sort cards rose during the morning, peaked in the afternoon and early evening and then fell away. Tasks with a heavier cognitive or memory load, such as mental arithmetic, had a different pattern and did not parallel the temperature rhythm. Not only the size but also the nature of both the cognitive and the memory load seemed to matter.

Most of the studies of fluctuations in circadian performance at that time were done, not unreasonably, during normal waking hours. Current thinking, based on more recent studies by Folkard among others, suggests that inter-task differences observed under normal conditions of sleep at night and being awake during the day fail to appear when subjects are tested at all circadian phases through a forced desynchrony protocol (for example the 28-hour days used by Czeisler). As a result (Carrier & Monk, 2000),

> performance efficiency on a specific task may decrease over the day because the amount of hours since awakening increases (homeostatic drive), because the input from the circadian timing system produces a less 'optimal' state to perform the task, or because of both these influences. In the same manner, performance efficiency may be stable over the day because the input from the circadian timing system exactly counterbalances the effects of increasing hours awake.

Performance rhythms are not simply due to circadian changes in either mood or physiology. Carrier & Monk (2000) point out:

The understanding of the mechanisms underlying different diurnal fluctuations during waking hours will require dissection of the individual effects of homeostatic and circadian influences on performance efficiency. This will not be a simple task since current research suggests that these processes vary with task parameters (e.g. cognitive load) and individual characteristics (e.g. age, chronotype, level of practice).

The demonstration of the circadian involvement in the sleep/wake cycle was of key importance in developing theories about sleep and alertness. Although sleep is the suspension of the activities of the awake state, it is more than the absence of being awake – it is not a default state but a regulated process. We need a certain amount of sleep every day. It is not surprising that there is an intimate connection between the circadian rhythm that controls an organism's temporal programming and the sleep process.

A Viennese neurologist with the wonderful name of Baron Constantin von Economo was the first person to discover the vital part played by the hypothalamus in sleep (von Economo, 1931). He studied the brains of individuals who had died from the epidemic of encephalitis lethargica. This epidemic has been popularised by Oliver Sacks in his book *Awakenings* (Sacks, 1991) and the film of the same name starring Robert De Niro and Robin Williams. Sacks came across an extraordinary group of survivors in a chronic care facility in the Bronx in the 1960s. Many of them had spent decades in strange, frozen states, like human statues, unable to initiate movement. He recognised these patients as survivors of the great pandemic of sleeping sickness that had swept the world from 1916 to 1927, killing over five million people.

Von Economo noted that damage to the front or anterior part of the hypothalamus caused insomnia, and suggested that this region contained a sleep-promoting area. By contrast, damage to the rear part of the hypothalamus caused sleepiness, and so he suggested that this region promoted wakefulness (Figure 11.3). He further hypothesised that lesions in this area could cause the disease we now know as narcolepsy.

Studies by other researchers in the 1940s identified another region of the brain that seemed to be involved in sleep. If regions in the basal

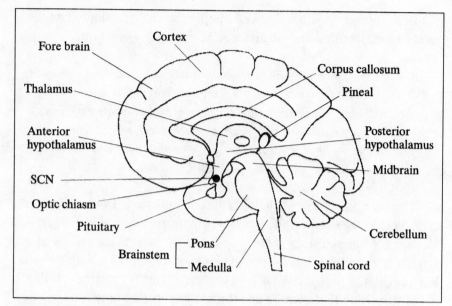

Figure 11.3. Diagram of a cross-section of a human brain, showing the relationship between the key structures involved in circadian timing and sleep.

brain (close to the spinal cord) were damaged, then sleep was induced, suggesting that there was a second wakefulness-promoting region located in the hindbrain. These early findings about the importance of the anterior and posterior hypothalamus, and of the brainstem, have subsequently turned out to be broadly correct.

The current model for sleep/wake regulation is shown in Figure 11.4, and the location of the brain nuclei and relevant neurotransmitters can be seen in Figures 11.5 and 11.6 (Pace-Schott & Hobson, 2002).

- The VLPO (ventrolateral preoptic nucleus) of the anterior hypothalamus promotes sleep. Neurones from this nucleus release GABA (gamma-amino butyric acid), an inhibitory neurotransmitter of the nervous system. The neurones project to and inhibit the activity of the nuclei of the ascending arousal system, and the lateral hypothalamus (LH).
- The ascending arousal system (AAS) of the brainstem and hypothalamus promotes wakefulness in the forebrain. Neurones from five regions in this complex (LDT, PPT, DR, LC, TMN) release several excitatory neurotransmitters. In the brainstem,

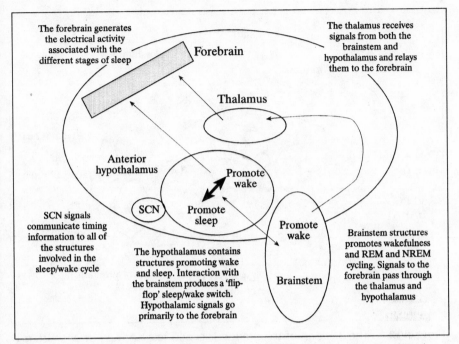

The forebrain generates the electrical activity associated with the different stages of sleep

Forebrain

The thalamus receives signals from both the brainstem and hypothalamus and relays them to the forebrain

Thalamus

Anterior hypothalamus

Promote wake

SCN signals communicate timing information to all of the structures involved in the sleep/wake cycle

SCN

Promote sleep

Promote wake

Brainstem structures promotes wakefulness and REM and NREM cycling. Signals to the forebrain pass through the thalamus and hypothalamus

The hypothalamus contains structures promoting wake and sleep. Interaction with the brainstem produces a 'flip-flop' sleep/wake switch. Hypothalamic signals go primarily to the forebrain

Brainstem

Figure 11.4. A simple schematic diagram of the interactions involved in the sleep/wake cycle.

neurones from the LDT (laterodorsal tegmental nuclei) and PPT (pedunculopontine tegmental nuclei) project to the thalamus, and from there to the forebrain. These two nuclei are responsible for the release of acetylcholine. The LC (locus coeruleus), also in the brainstem, has neurones that project to the forebrain and release noradrenaline. The DR (dorsal raphe nucleus) of the brainstem has neurones that project to the forebrain and release serotonin. In the hypothalamus, neurones from the TMN (tuberomammillary nucleus) project to the forebrain and release histamine.

- The lateral hypothalamus also promotes wakefulness. Neurones from this nucleus release orexin (also called hypocretin), a very recently discovered neuropeptide. Neurones project to the nuclei of the ascending arousal system, the forebrain, as well as the VLPO.
- The NREM/REM oscillator is a cluster of five separate nuclei in the brainstem that provides the switch between NREM and REM sleep. Three of these nuclei, the LDT, PPT and BRF (brainstem

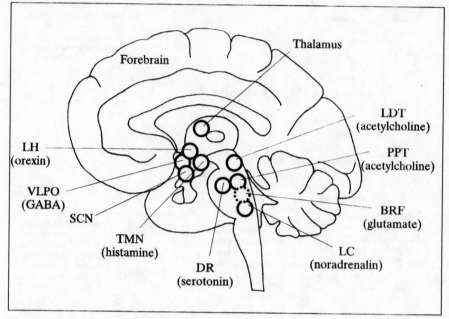

Figure 11.5. The location of the brain nuclei and relevant neurotransmitters involved in sleep generation. BRF, brainstem reticular formation; DR, dorsal raphe nucleus; GABA, gamma-amino butyric acid; LC, locus coeruleus; LDT, laterodorsal tegmental nuclei; LH, lateral hypothalamus; PPT, pedunculopontine tegmental nuclei; SCN, suprachiasmatic nuclei; TMN, tuberomammillary nucleus); VLPO, ventrolateral preoptic nucleus.

reticular formation), are interconnected and excite the activity of both themselves and the two other nuclei involved in the NREM/REM oscillator. The LC and DR form a second functional unit. The LC and DR are interconnected and inhibit the activity of both themselves and the LDT, PPT and BRF functional unit. This reciprocal set of interactions generates a flip-flop switch that produces a roughly 90-minute oscillation in NREM and REM sleep.

- The suprachiasmatic nuclei (SCN) regulate the various sleep structures of the brain either directly by neural or chemical outputs, or indirectly by the release of the pineal hormone melatonin. Melatonin is high throughout the night, and when administered has been shown to increase the propensity for sleep in humans.

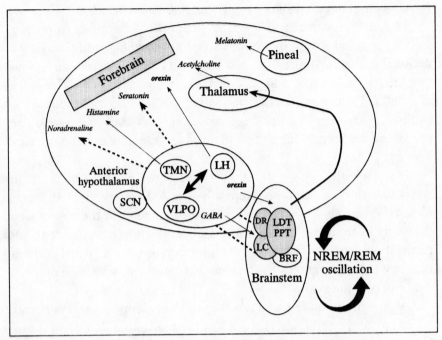

Figure 11.6. Composite diagram (not to scale) illustrating the major interactions of the brain structures involved in the generation of sleep and wake. The circadian modulation of sleep originates within the suprachiasmatic nuclei (SCN), and is likely to involve the rhythmic production of melatonin from the pineal. The VLPO (ventrolateral preoptic nucleus) is in direct contact with the lateral hypothalamus (LH) and the various nuclei that constitute the ascending arousal system (LDT, laterodorsal tegmental nuclei; PPT, pedunculopontine tegmental nuclei; LC, locus coeruleus; DR, dorsal raphe nucleus; TMN, tuberomammillary nucleus), and promotes sleep by releasing the inhibiting neurotransmitter GABA. The awake state is promoted by the release of orexin from the LH and the activation of the ascending arousal system. The switch from NREM/REM sleep is achieved by the interaction of the LDT, PPT, BRF (brainstem reticular formation), LC and DR.

Sleep interacts with other hypothalamic functions that are under circadian control. Sleep deprivation alters hormone release, increases body temperature, stimulates appetite and activates the sympathetic nervous system. Understanding this interaction between brain structures is one of the biggest challenges in neuroscience and is part of the bigger programme of determining how all of the brain's systems together create the collection of behaviours we call the 'self' (Saper *et al.*, 2001).

Despite the complexities of the sleep/wake system, current treatments for sleep disorders act on a limited number of pathways. The detailed understanding of the system and its circadian component suggest that a new generation of drugs will become available (Mignot *et al.*, 2002). Each year, tens of billions of dollars are spent globally on hypnotics, and many of the drugs have long half-lives and a wide variety of side effects. Hence the increasing research effort into sleep activity.

It is likely that new generations of drugs will be developed that act selectively to help promote sleep, as is needed for people with insomnias, or wakening, for narcoleptics. As the estimate for the cost of sleep disorders in the USA alone in terms of production loss, accidents and medication is of the order of $40 billion a year, it is a virtual certainty that new, more targeted, pharmaceutical interventions for sleep disorders will become available.

Although the quality and quantity of their sleep has always been of major concern to athletes, there is now the realisation that when the differences between success and failure in high-class sporting activities are so small, circadian variation can have a significant effect. During the mid-1980s, Britain's quartet of world-beating middle-distance runners, Seb Coe, Steve Ovett, Steve Cram and David Moorcroft, repeatedly broke many world records, but always during the evening. It could be that the evening was the best time for middle-distance running because that is when the body temperature is highest. A number of measures of possible relevance to athletic performance can vary with the temperature rhythm. For example, muscle strength, isometric strength of the knee extensors, whole-body flexibility and self-selected work rate are all rhythmic with peaks around 6.00 to 8.00 p.m., which is also the time of peak body temperature. Or it could have been that the demands of the promoters and the TV audiences meant that world record attempts were only scheduled for the evening.

Controlling for all the variables that can affect an athlete's performance is something of an experimental nightmare. Thomas Reilly and his co-workers at Liverpool John Moores University studied the performance of swimmers, because at least it is possible to control for weather and ambient temperature with the added bonus that indoors there is no wind (Atkinson & Reilly, 1996). There was a significant linear improve-

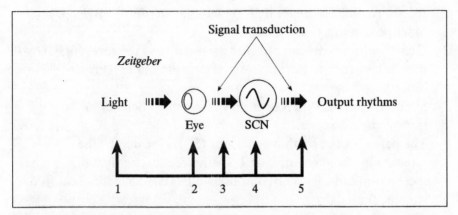

Figure 11.7. Modifying individual chronotypes by intervention at one or a combination of five possible points of variance.

ment throughout the day: the time for 100 metres was 2.7 seconds faster at 10.00 p.m. than at 6.30 a.m. Quite why there should be such an improvement is not yet known, but perhaps the 0.6–0.8°C rise in core body temperature in the evening has a small but significant effect on nerve conductance speed or muscle efficiency.

Being good at a sport is not just a case of natural ability. Talent is needed but so is dedication in training. Would-be Olympians spend hours on the running track, the ice rink and in the pool, often early in the morning. Given that there are morning and evening chronotypes, it is a moot point as to how many would-be great athletes have given up because of their training schedules. One study of chronotypes found that golfers, who competed at different times of the day, scored higher for morningness than water polo players, who performed mostly in the evenings (Rossi *et al.*, 1983). Reilly recommends that such individual differences in chronotype should be recognised by coaches when scheduling training sessions. 'For example, an extreme evening type swimmer may find that attending early morning training sessions is so difficult that he or she quits the sport, even though he or she performs well in evening competition' (Atkinson & Reilly, 1996).

We all have an individual chronotype or time signature. If we recall the modified simple model of circadian organisation, there are at least five points where variance can occur (Figure 11.7).

1 The strength of the *Zeitgeber* will vary. For example, a farmer receives a much stronger light signal than an office worker or an older person in a nursing home.

2   Genetic differences can affect the sensitivity of the receptor, so two people may receive the same signal strength but perceive it differently.

3   Similarly, genetic differences may affect the efficiency of the input transduction cascade.

4   The period of the rhythm generator (SCN) created by the interaction of the negative and positive feedback drives of the clock mechanism can vary. Different individuals can have different free-running periods.

5   The control of the output signal can vary, again depending on genetic factors.

The combination of these five points of variance affects our mental and physical outputs so that some of us are better at doing certain things at one time and others at another. It is impossible to say how much more productive we would be if we timed our activities so that they were in harmony with our time signatures, but if it makes the difference between a gold medal and a bronze, and peace and war, then it is probably worth it.

1 2

SAD SHIFTS

When you go to Europe, the soul takes about three days longer to get there.
C. DUNLOP AND J. CORTAZAR, IN *LOS AUTONAUTAS DE LA COSMOPISTA O*
UN VIAJE ATEMPORAL, PARIS-MARSELLA MUCHNIK, BUENOS AIRES 1983
(GOLOMBEK & YANNIELLI, 1996)

In late afternoon, the European squirrel burrows into its underground nest in the dry grassland steppe of the Hungarian puszta and does not emerge till mid-morning. This photic environment is similar to that which we humans have created for ourselves. Many of us, like the squirrel, seldom see dawn or dusk, particularly in the winter when we trudge off to our work before the sun has risen and travel home in darkness. Yet the squirrel's biological clock somehow manages to use whatever daytime outdoor light it sees to synchronise activity – and reproduction and hibernation – to the changing daylength of the seasons. We are not so fortunate with our artificial photic environment.

We use artificial light to extend our period of wakefulness and activity into the evening hours, and a short sleep schedule which consolidates sleep efficiently at night. It probably began back in prehistoric times when our ancestors discovered fire. Nowadays, we generally receive less light in total during the day than the European squirrel. Even in somewhere as sunny as San Diego, people in summer receive approximately two hours of natural light, or less, a day. Among healthy older people it is much less – only 75 minutes in every 24-

hour period for men and just 20 minutes for women (Cole *et al.*, 1995).

According to Richard Stevens and Mark Rea (Stevens & Rea, 2001),

Daylight does not effectively penetrate building spaces deeper than approximately five metres and window blinds commonly keep out direct sunlight. People rarely look directly at an electric light or the sun because these sources commonly evoke a photophobic, discomforting response. People working indoors view objects and surfaces illuminated at between 50 and 500 lux. These light levels are significantly dimmer than those experienced by those who work outdoors, such as construction and agricultural workers who view objects and surfaces illuminated by the sun and daylight at about 5000 to 100,000 lux.

While humans have increasingly insulated themselves from the natural cycles of light and darkness, the human circadian pacemaker has conserved a capacity to detect seasonal changes in daylength. For many of us this means there is a price to pay for our artificial environment. About 3 per cent of people in the UK suffer from Seasonal Affective Disorder (SAD), a mood disorder involving a recurring autumn or winter depression. The worst months are October through to April. Tromsø in Norway is at latitude 69° N and the sun does not rise above the horizon from 20 November until 20 January. Midwinter blues are estimated to affect 27 per cent of the population (J. Edvardsen, University of Tromsø, personal communication). The condition was unknowingly described by the Arctic explorer Frederick A. Cook in his journal on 16 May 1898 (Wurtman & Wurtman, 1989):

The winter and the darkness have slowly but steadily settled over us ... The curtain of blackness which has fallen over the outer world of icy desolation has also descended upon the inner world of our souls. Around the tables men sitting about sad and dejected, lost in dreams of melancholy from which, now and then, one arouses with an empty attempt at enthusiasm.

For people with full-blown SAD the disorder can be life-shattering. Seriously depressed, they withdraw from social activities and in the most severe cases may commit suicide. People with SAD crave carbohydrates and often gain 5–10 kilograms. When spring comes around, they begin to feel better. As the days get longer, their mood improves, they have renewed energy and may lose the weight they gained in a way analogous to hibernating animals.

It could be that SAD is a vestigial trait that was of benefit to early hominids who were migrating away from equatorial Africa to higher and more seasonal latitudes. As winter drew in and food and light were significantly reduced, levels of the mood-altering neurohormones such as serotonin fell, the hominids in high latitudes may have gone into a state of semi-somnolescence. The noted sluggishness of SAD sufferers is analogous to the curtailment of energy expenditure in hibernating animals. Similarly, the increased amount of food and light of springtime corresponds with higher levels of serotonin, sparking the appetite for food and sex (Magnusson & Boivin, 2003). In modern times, because humans cannot afford to slow down in the wintertime, SAD serves as a hindrance and not the benefit it may once have been.

In our artificial world and its absence of daily sunlight, we use a range of cues, most notably the alarm clock, to entrain us to the solar cycle. But when we sleep at irregular times, fly across time zones or work shifts, we are reminded that in the 24-hour society we have created it is not only during the winter that our bodies are fighting against a deeper rhythm that Alexander Chizhevsky described as an 'echo of the sun' (Halberg *et al.*, 2001).

The problems come after a long flight or a series of shifts when, for example, the clock says it is 9.00 a.m. and your body clock is reading 4.00 a.m. After a wretched day most of us can get our rhythms back into synchronisation after a night's sleep. Getting over jet-lag can take longer. Almost everyone who has flown across a few time zones has felt at least some of the symptoms of jet-lag: fatigue, an inability to fall asleep, body aches, digestive problems and disorientation. In their advice to travellers, British Airways warns that the effects can be severe, with decision-making ability downgraded by up to 50 per cent, communication skills by 30 per cent, memory by 20 per cent and attention by 75 per cent.

The symptoms of jet-lag often seem to be worse on the second and third days after the time shift, compounded as they are by poor sleep. This could be because different body rhythms (for example, temperature, endocrine, heart dynamics, gastrointestinal, sleep/wake) adjust at different rates. On the first day, all rhythms are equally out of phase. On the second and third days, rhythms are out of synchrony with each other, compounding the feeling of jet-lag. This circadian disharmony can be compared to an orchestra that starts with one conductor, playing in harmony, but when another conductor appears with a different beat there is cacophony as the pieces convert at different rates to the new maestro. Each musician in the orchestra can play his or her own part in the music, but if they do not play it at the right time, the result is not music.

Normally, the light and dark periods, temperature, humidity and social interactions entrain our bodies to a 24-hour cycle. When these variables are mixed up – such as when one flies from San Francisco to London – body processes don't catch up immediately. It can take up to a day for each time zone travelled for most people to adjust to the environmental stimuli of a new place (Appendix II).

The symptoms are more prolonged going from west to east. One study of US Army soldiers transferring from the USA to Germany found that it took eight days to adjust to time and environmental changes, whereas soldiers flying home to the USA took just three days (Wright *et al.*, 1983).

Bill Schwarz, whom we met in Chapter 5 when he was working at his day job as a neurology professor at the University of Massachussetts, is a passionate baseball fan. With the help of colleagues he has analysed the records of West Coast baseball teams playing away games on the East Coast, and vice versa. The results are what every West Coast sports fan instinctively knew. Their teams are systematically robbed. The home team scored 1.24 more runs each game when the visiting team had just completed West-to-East Coast travel. The home games played in the west showed no statistically significant difference in runs per game (Recht *et al.*, 1995).

Why is it harder to adapt to travelling east? Schwartz points to the fact that, devoid of stimulus, the body develops rhythms slightly longer than 24 hours. He suggests that it might be easier for the body to adapt to a longer day, such as when an airline passenger gains three hours on a trip from New York to Los Angeles, than to a shorter day when travelling eastward. 'When you fly from east to west, you're lengthening the day and going in the natural direction the (internal) clock wants to go', Schwartz said. 'When you're flying from west to east, you're compressing the day' (Recht *et al.*, 1995).

One way to deal with jet-lag is to carry a pair of sunglasses. Light shifts the biological clock, but light does not have the same shifting effect on the clock at different times of the day. Light during the early night delays the clock, so we would start to sleep later the next day. Light during the late night, after the temperature minimum at 4.00 a.m., advances the clock so that we would start our sleep earlier the next night. Because light at different times of the night can either advance or delay the clock, it is essential when crossing multiple time zones to make sure that light exposure in the new time zone moves the clock in the right direction (Appendix II).

The problems associated with jet-lag are compounded because whereas the SCN adjusts quickly, the peripheral clocks in various organs such as the liver take much longer. Michael Menaker and his colleagues put groups of rats through six-hour phase shifts to mimic a transatlantic flight, then sacrificed them at various times afterwards and looked for oscillation peaks of clock genes in tissues. The SCN adjusted almost immediately, but peripheral structures' phases shifted very slowly. The liver, for example, took a particularly long two weeks to catch up with the rest of the system (Yamazaki *et al.*, 2000).

These results support the long-held hypothesis explaining the symptoms of jet-lag sufferers and night-shift workers who have a hard time adjusting to phase changes: the human body is experiencing internal desynchrony. According to Menaker, if scientists could identify the signals that put the SCN in touch with peripheral oscillators, they might be able to fashion a treatment by manipulating those signals (Yamazaki *et al.*, 2000).

The effects of occasional jet-lag are temporary, but continual flying backwards and forwards across time zones has long-term effects. Kwangwook Cho scanned the brains of female flight cabin staff with at least five years' flying experience and discovered that those who made regular journeys across several time zones demonstrated evidence of impaired thinking ability. Cho explained: 'Cortisol levels in cabin crew after repeated exposure to jet lag were significantly higher than after short distance flights. And the higher cortisol levels were associated with cognitive defects' (Cho *et al.*, 2000).

The stunning increase in trans-meridian flights since the first commercial jet flight by a DeHavilland Comet in 1949 is just one symptom of our globalising '24/7' society. In the developed world, regular shift working is still far less common than it was in the early parts of the twentieth century, but the same forces are pushing us to work to the beat of a planetary clock. Upwards of 20 per cent of the population of employment age work at least some of the time outside the 7.00 a.m. to 7.00 p.m. day (Kreitzman, 1999). This is not without consequence, as Josephine Arendt of Surrey University has pointed out (Rajaratnam & Arendt, 2001):

> *Because of their rapidly changing and conflicting light–dark exposure and activity–rest behaviour, shiftworkers can have symptoms similar to those of jetlag. Although travellers normally adapt to the new time zone, shiftworkers usually live out of phase with local time cues.*

Shift workers suffer from the double bind that even if they try and get their full sleep quotient, they may not be able to sleep when they want to. Coming on to the evening or night shift can require a delay in the normal sleeping time of as much as 3–12 hours. The shift worker may wish to sleep, but the circadian rhythms may be preventing it. Harold Thomas, a physician with long experience of hospital emergency rooms, has drawn attention to some of the long-term biological and social problems associated with shift working (Whitehead *et al.*, 1992):

> *Physical problems include an incidence of peptic ulcer disease eight times that of the normal population. Cardiovascular mortality has also been noted to be increased among shift workers. One author estimates that the risk of working rotating shifts approaches that of smoking one pack of cigarettes per day. Other physical problems include chronic fatigue, excessive sleepiness, and difficulty sleeping. Part of the social toll on those who must work rotating shifts is reflected in an increased divorce rate. Shift workers have higher rates of substance abuse and depression and are much more likely to view their jobs as extremely stressful.*

Despite this, there are some individuals who prefer to work permanently on night shifts. It could be that they are extreme owls, who find it easier to be awake and alert at night, or it may be a more mundane reason such as higher pay at night. One Nobel Prize-winning chemist was known not only for the quality of the syntheses he devised but also for the sheer volume of work he produced. It was said that he had an unhappy marriage and preferred to stay in the laboratory at night as opposed to being at home.

Although there is some self-selection by those who prefer working at night, the avoidance of it by those who find themselves intolerant of night work makes it difficult to pin down the epidemiological consequences of shift patterns. But there are few dissenters to the assertion that shift work is bad for you. Even if a night worker has slept reasonably well, he or she is no more alert between 2.00 a.m. and 8.00 a.m. than a day worker who for two nights in a row has only slept four hours a night. This is compounded by constantly changing shift patterns that play havoc with the circadian clock system. The various symptoms associated with shift schedules have been lumped together under the generic term shift-lag, and an industry has developed advising employers and employees how best to counteract the effects of working shifts.

Schedules can be classified by three factors: the length of the shift; the speed of the change in schedule, for example a week on days rotating with a week on nights; and the direction, for example clockwise would be day shifts, followed by evenings, then nights. There is no consensus as to which shift pattern is best, because even the view that forward rotation is better than the reverse has not been conclusively shown.

Simon Folkard considers (Folkard, 2000) that

> the root of the problem is that man is primarily a day-active animal. There is the parallel of man being mainly a land animal, not designed for living in water. Similarly we are not naturally night-active. A huge number of body rhythms from sleepiness/wakefulness to digestive enzymes, work to a basic 24-hour day-active rhythm. These rhythms do not adjust immediately to a night-active pattern.

As if working on the night shift were not in itself a sufficient hazard, workers are also vulnerable when driving home after a night shift, especially on quiet monotonous roads. There is a 50 per cent increase in the risk of a single-vehicle crash at 3.00 a.m. after four successive night shifts, and the risk can be even higher between 7.00 and 8.00 a.m. Because of the profound influence of the circadian rhythm, sleep-related crashes tend to occur after even a short period of driving. There is a more modest rise in accidents at 2.00 to 4.00 p.m. These are distinct from the other more general peaks in road crashes that occur in the rush hours of 8.00 to 9.00 a.m. and 5.00 to 6.00 p.m. Corrected for traffic density, the relative risk of an accident occurring is 20 times higher at 6.00

a.m. than at 10.00 p.m. (National Commission on Sleep Disorders Research, 1992).

Many of the big maritime and industrial accidents have happened at night. The first distress call from the *Titanic* went out at 12.15 a.m. local time. The nuclear accident at Three Mile Island began at 4.00 a.m., Chernobyl at 1.23 a.m. and the explosion at the Union Carbide plant in Bhopal, India, at 12.15 a.m. If that were not enough, shortly after midnight on 24 March 1989 the T/V *Exxon Valdez* ran aground on Bligh Reef in Prince William Sound, Alaska, spilling almost 11 million gallons of North Slope Crude oil. The ferry *Estonia* sank in less than 20 minutes in the Baltic Sea at 1.50 a.m. on 28 September 1994, with the loss of 852 people.

Not all the blame can be laid on fatigue or messed-up circadian rhythms. If a boat is going to hit an iceberg, it is likely to be at night when the iceberg cannot be seen. Likewise, if a group of nuclear plant technicians at Chernobyl who have been on duty for 13 hours start being macho and make a series of poor judgements, they are more likely to do it at night than during the busy day. The result of that was damage of about £200 billion and the deaths of several thousand people.

The *Exxon Valdez* disaster happened at night, but not because the captain had been drinking, as is often supposed. He was not on duty at the time of the accident. The problem seems to have been a shift change that occurred just before the accident. The lookout was helping the harbour pilot get off the boat and an inexperienced third mate was at the helm. The cause of the accident was not simply tiredness but a combination of poor shift-work scheduling, human error, excessive overtime, crew fatigue, inappropriate sleep schedules, inadequate shift changes, alcohol, insufficient training and reduced crew sizes. That little lot resulted in oil spill devastating over 1,300 square miles.

Folkard has pointed out, however, that fatigue or slow responses were at the very least partly responsible for all the above incidents and there is no getting away from the fact that night work can be a problem (personal communication). Those who work at night perform less well and make more errors. In Britain, sleep-related workplace accidents are reckoned to cost at least £114 million a year. The risk of injury on the night shift is over 20 per cent higher than on the day shift. The key NASA officials involved in the *Challenger* disaster made the decision to go ahead after working for 24 hours straight and having had only two to three hours of sleep the night before. Their error of judgement cost the lives of seven astronauts and nearly killed the US space programme (Moore-Ede, 1993).

Space is the most extreme shift-working environment. The crews on space missions sleep poorly. On some space shuttle missions up to half the crew take sleeping pills, and, overall, nearly half of all medication used in orbit is intended to help astronauts sleep. Even so, space travellers average about two hours' sleep less each night in space than they do on the ground.

NASA sometimes deliberately shifts the astronauts' body clocks before sending them into space, making sure that their biological day coincides with the crucial period of launch. Effecting the shift is easy: astronauts are exposed to high-intensity light at key times for 3–10 days before lift-off. Once in orbit the biological clocks of astronauts might need to be adjusted further to align with another critical time – the moment of landing. They do this by requiring the crew to wake up earlier and earlier each day.

Light is used in therapeutic attempts to counteract the effects of jet-lag, shift-lag and various sleep disorders. It is usually the first line of treatment for SAD. Sometimes banks of lights are used to administer at least 45 minutes of simulated daylight – about 2,500 lux, or the equivalent of being outside on a cloudy, grey day. Timing and light intensity are the keys. There seems to be a 'critical period' between 3.30 and 8.00 a.m. when the retina must receive the light for the treatment to work. Jeffrey Rausch was part of the research team in California in the early 1980s that hit on the link between morning light and SAD. These researchers looked for causes and treatments associated with light and considered the amount of light exposure that sufferers had in their office and at home, and whether they slept on the northeast side of the house with the curtains closed and did not receive light until 8.00 or 9.00 a.m.

As well as light therapy, melatonin is the chemical treatment of choice for shifting the human circadian clock and so circumventing some of the problems associated with, say, jet-lag. Well over 20 million Americans have been estimated to have taken melatonin for one reason or another, even though the naturally occurring hormone is classified as a food supplement and so has not been through rigorous clinical trials. As it is classified as a natural hormone it cannot be patented so there is no commercial interest in running proper trials. Instead there is a great uncontrolled trial by the millions who take it.

Apart from its effects on biological timing it is regarded as something of a wonder drug and there have been claims that that it can reduce cancer, high blood pressure, Alzheimer's disease, AIDS and coronary heart disease as well as improving sleep, sexual vitality and longevity. At the height of the hype in the mid-1990s, melatonin even made the front cover of *Newsweek* magazine.

Not all the claims are outlandish. University of Connecticut epidemiologist Richard Stevens points to a study that found elevated breast cancer rates among Finnish flight attendants, noting that the incidence is too high to be accounted for solely by increased radiation exposure. Disruption of circadian rhythms might well be a causative factor in these cancers, he says. It is also the case that alcohol disrupts sleep, which in turn could suppress melatonin, perhaps explaining why excessive alcohol consumption increases breast cancer risk (Batt, 2000).

Taken during the day (when it is not naturally produced) melatonin can induce sleepiness and impairment of performance. So does it cause sleep? Numerous experiments have shown that people can become sleepy 30–120 minutes after taking melatonin. But by no means everyone feels sleepy after taking it. And, not surprisingly, it does not have a hypnotic effect in the rat. The rat is nocturnal and so sleeps during the day and is active at night. But the rat releases melatonin at night when it is active. So the rat does not use melatonin for sleep, and melatonin does not make rats sleepy – or other nocturnal animals for that matter.

Melatonin is critically important in driving the mammalian photoperiodic response (Chapter 9). Melatonin can also affect sleep in some diurnal mammals, but uniquely in mammals melatonin was thought not to be involved in the circadian system. Removal of the pineal had little or no effect on the ability of mammals to entrain to a light/dark cycle. This view was challenged in experiments by Jenny Redman and co-workers at La Trobe University in Australia. They showed that daily injections of melatonin at the same time each day would eventually entrain the free-running activity rhythms of rats in constant darkness. The rats would start their activity at about the same time as melatonin was being given (Redman *et al.*, 1983). Subsequent experiments by this group and others have shown that, if appropriately timed, melatonin can speed up resynchronisation of rodent daily rhythms after a shift in time.

Blind people are generally unable to synchronise their sleep/wake and activity rhythms to the solar day because they cannot be entrained by light. This is particularly true of patients who have had their eyes surgically removed as a consequence of a tumour or other illness. Although only a few cases have been studied, a daily dose of melatonin at the desired bedtime seems to help about a third of patients to synchronise their sleep/wake cycle to the 24-hour period (Arendt *et al.*, 1988). Jo Arendt, who has done much of the work with blind people, was the first to investigate the effect of daily melatonin administration on sighted humans' biological rhythms. Compared with a placebo, melatonin treatment produced an advance in the timing of its own internal rhythm, an advance in the timing of the hormone prolactin rhythm and an advance in the timing of subjective evening tiredness. But Arendt also made it clear that the timing of the melatonin administration was critical. She says, 'Taking melatonin at the wrong time may disrupt biological rhythms. Melatonin-induced sleepiness may impair driving or job performance' (Arendt *et al.*, 1995).

However, she and others have shown that melatonin can be used to help alleviate various circadian disorders. Self-rated jet-lag can be reduced by about half, as long as the melatonin dose is appropriately timed. The improvement increases with the number of time zones traversed and is greater going east than west. As yet there have been very few instances of melatonin use in overcoming the problems inherent in shift-work patterns.

Melatonin is commercially available in concentrated form in tablets at far higher levels than would be found in the blood. The circulating daytime serum levels of melatonin in healthy adults do not normally exceed 20 nanograms (billionths of a gram) per litre, whereas the range of the night-time values may be about 20–170 nanograms per litre. There is considerable variation in the daily amount of melatonin produced between people. The long-term effects of high levels of melatonin administration are unknown. They may not necessarily be advantageous.

13

TIME TO TAKE YOUR MEDICINE

Just as you prescribe a dose for every drug, you must prescribe a timing for every drug.

FRANZ HALBERG, DIRECTOR OF THE CHRONOBIOLOGY LABORATORY AT
THE UNIVERSITY OF MINNESOTA (KRUSZELNICKI, 1998)

Alexander the Great died from it. So did Oliver Cromwell. Until modern times, malaria epidemics were common in Europe. Malaria ('bad air') was rife in the Campagna region of Italy – the plain surrounding Rome – and many of the cardinals gathered at the Vatican in the seventeenth century would die of the disease while choosing a new Pope. The standard remedy of the time was based on bleeding, but the Jesuits vigorously pushed the New World remedy of extracts of cinchona bark. Anti-Catholic prejudice was so strong that many Protestants refused to use it, including Cromwell, who died because he would not take a remedy endorsed by Rome. But cinchona extracts gradually became the treatment of choice, and in 1820 two French scientists isolated quinine as the active ingredient.

Today malaria is a disease of the tropics, where it affects between 300 million and 500 million people, of whom about two million, mainly children, die (WHO, 1992). When an infected female *Anopheles* mosquito bites a human to drink blood, the *Plasmodium* parasites she is harbouring are injected into the bloodstream. The parasites infect the liver and then the red blood cells. The blood cells rupture and billions of parasites are simultaneously released to invade new erythrocytes and in the process dump high concentrations of toxins into the bloodstream. The host recognises this poisoning and responds by releasing its own chemicals, especially tissue necrosis factor (TNF). This chemical turns up the body's thermostat and changes the set point, which gives the infected host the sensation of extreme coldness. The induced shivering helps raise the body temperature to 39–40°C. Nausea and vomiting follow the rapid fever, only to be followed in turn by the 'hot phase', with its sensations of extreme heat, headache and delirium. Aided by copious sweating, the fever subsides over a period, the whole cycle being termed a 'paroxysm'. Each time, the recurrent fever is a response to the new wave of infection.

Plasmodium has evolved effective resistance to quinine-based drugs and there is now some urgency in the need to develop effective malaria interventions. Sterile mosquitoes have been released in huge numbers in the hope of reducing mosquito populations. Vaccines are being developed, but genetic diversity and antigenic variation makes effective development very difficult, though the publication of the *Plasmodium* genome in autumn 2002 should intensify the research.

What makes malaria the quintessential chronobiologic disease is the rhythmic nature of the illness. Depending on the *Plasmodium* species, two, three or four days after the first rupture of red blood cells, another wave of erythrocytes rupture and the parasites are synchronously released. But if the *Plasmodium* parasite is grown in erythrocytes in culture, the synchronicity of release is rapidly lost. Further, when the pineal is removed from melatonin-producing mice infected with *Plasmodium*, there is also a loss of synchronicity of erythrocyte rupture. These experiments suggest that the timing signal comes from the host. Essentially, *Plasmodium* 'captures' the host's timing mechanism and turns it back on the host so as to increase its infectivity.

Celia Garcia and her colleagues at the University of São Paolo have now shown that host melatonin triggers a cascade of events that lead to a release of calcium ions from the *Plasmodium* parasite and that this calcium signal is the synchronising trigger (Garcia *et al.*, 2001). If the *Plasmodium* melatonin receptor can be blocked, then the resulting desynchronisation of the parasite's release can be combined with drugs that act on specific stages in the cycle and so improve their therapeutic efficacy.

The complex life cycle of *Plasmodium* and its close interaction with the host's circadian timing system is a sophisticated example of the therapeutic possibilities inherent in developing an understanding of the biological role of such rhythms. The same detailed analysis of the timing aspects of a wide range of diseases, including cancers, so as to optimise surgical and pharmaceutical intervention goes under the name of chronotherapy.

Many physiological indicators have huge amplitudes. Melatonin levels rise during the night and are virtually non-existent during the day, whereas cortisol production is at a low point at the start of the sleep episode and rises during the morning. This amplitude is critical in diagnosis rather than the usual single-point analysis. Whether somebody has a 'normally' low melatonin level when measured during the day is meaningless unless the night-time levels are also known, because the subject could have abnormally low melatonin across the 24 hours. Likewise, women with anorexia have 'normal' prolactin levels measured during the day but do not show the night-time surge in the hormone's level.

One of the standard tests for asthmatics is to measure airway function, which is usually higher in the afternoon than in the morning. So an early-morning appointment with the doctor may confirm the severity of the condition whereas the same person seeing the same doctor later in the day may well have a different result. Same person, same doctor, same disease – different time.

It seems common sense to deliver a drug at the time when it will be most effective (Elliott, 2001). But most medication is not prescribed to intersect with the changing physiology and biochemistry of the patient. Instead, drugs are taken at regular intervals, largely with the aim of maximising compliance rather than effectiveness. Most prescribing is done in the hope of maintaining a stable drug level and trusting that the drug manufacturers have balanced therapeutic dosages with toxicity levels so that the drug works. Whether it does or not is a moot point.

Rats or mice are commonly used as experimental animals in toxicity testing of drugs and additives. But rats and mice are nocturnal animals. Robert Burns has described a scenario indicating some of the problems associated with working on resting animals during their diurnal phase and transferring the data to the diurnally active human. He says (Burns, 2000):

A chemical food additive is tested in a mouse model for carcinogenicity. The test is done when the susceptible biochemical event (e.g. rate of DNA synthesis) happens to be at its lowest activity, i.e. at mid-day. No cancers are produced. The data are accurate: exposure to the chemical at mid-day did not result in any cancers. The data are used to classify the chemical as non-carcinogenic and it is approved for use as a food additive. However, if the same dosage had been tested at midnight, a time of maximal susceptibility of the biochemical event involved, 40% of the animals would have developed liver cancer and the compound would be classified as carcinogenic and unsafe.

The goal should be to balance the circadian variation in pharmacokinetics (what the body does to the drug) with the circadian variation in pharmacodynamics (what the drug does to the body). According to Georg Bjarnason of the University of Toronto, by 1999 a statistically significant circadian variation had been documented for the pharmacokinetics of over 100 drugs, including many antineoplastic agents used in cancer treatment (Bjarnason *et al.*, 1999). Whether a drug gets distributed to the site of action at a sufficiently high concentration to do any good depends on the rate of absorption, the rate at which it is metabolised and the rate at which it is excreted. All these are subject to circadian variation.

Promising drugs to treat arthritis in older people have failed because excretion was slower than it was in younger people, who, in the manner of things, were the ones on whom the drugs were tested for toxicity. As a result, the drug hung around in the body for too long and reached toxic levels in older people. Timing the dose more carefully might have worked.

Yet doctors have done little to acknowledge the importance of timing in clinical practice, despite pioneers such as Bill Hrushesky, an American oncologist at the University of South Carolina School of Medicine, who has been railing about it for years. He points out (personal communication) that

For more than 5,000 years, traditional Chinese medicine has recognised the concept that the dose cannot possibly be isolated from the concept of timing. An entirely different dose of a certain drug or an entirely different drug is routinely prescribed for a single ailment at different times of the day, different times within the week, menstrual cycle, or year.

There is no shortage of evidence that there is something in the Chinese practice. Nearly 20 years ago, Hrushesky published a paper in *Science* reporting an experiment in which he switched the timing of chemotherapy in 31 women who had ovarian cancer. In his trial Hrushesky divided the women into two groups. Each group received two standard cancer drugs, adriamycin and cisplatin, but one group received adriamycin at 6.00 a.m. and cisplatin at 6.00 p.m., while this daily schedule was inverted in the other group. He found that the women on the adriamycin 6.00 a.m./cisplatin 6.00 p.m. schedule developed roughly half the side effects of the women on the other, with fewer treatment delays. There was less hair loss, less nerve damage, less kidney damage, less bleeding, fewer transfusions: 'Every toxicity was markedly diminished several-fold simply depending on what time of day the drugs were given' (Hrushesky, 1985).

The fundamental challenge in oncology is to kill the tumour without killing the patient. The drugs used in chemotherapy are highly toxic and can damage vital organs such as the heart and kidneys, and radiation treatment also can have considerable and often damaging side effects. Further, it is extremely difficult to destroy every cancer cell. Only a few need to survive to proliferate and cause the reappearance of the disease. Although the hair loss that is characteristic of these therapies is cosmetically and psychologically unpleasant, it is minor compared with the nausea, vomiting, bone marrow suppression, diarrhoea, debilitation and the peripheral sensory neuropathy (the inability to feel with hands or feet) that accompany such treatment regimes. Those who followed the plight of the fictional Dr Green as he battled with the cerebral tumour in the TV series *ER* will recall that he decided that the treatment was worse than the disease.

The strategic aim of most of the non-surgical therapies is to attack the cancer cell's replication machinery. This includes not just the nucleic acid but also the cell's architectural components that are essential to division, such as the cytoskeleton and the microtubules present in the cytoplasm. Another line of therapeutic attack is on the protein-synthesising mechanism in the cancer cell. The therapeutic art is to devise ways and means of delivering a sufficiently potent dose of a highly toxic chemical or radiation without killing too many normal cells at the same time.

Cells multiply through cell division, but before a cell can divide it has to grow in size, duplicate its chromosomes and separate the chromosomes for distribution between the two daughter cells. The 2001 Nobel Prize was awarded to Leland Hartwell, Paul Nurse and Tim Hunt for their work in elucidating the cell cycle and determining the key checkpoints and the nature of the proteins that drive the process. The cycle consists of several phases. In the first phase (G1) the cell grows. When it has reached its appropriate size it enters the next phase, DNA synthesis (S), in which the chromosomes are duplicated. During the next phase (G2) the cell prepares for division. In mitosis (M) the chromosomes separate, and the cell divides into two daughter cells. Through this mechanism the daughter cells receive identical sets of chromosomes. After division, the cells are back in G1 and the cell cycle is completed (McDonald & El-Deiry, 2000).

After a sufficient number of new cells have been generated, perhaps to replace old cells or heal a wound, the cell-cycle-controlling proteins are switched off and cell division ceases. In a cancer cell the machinery goes wrong and the cell keeps on dividing. The change from normal to cancerous function can be caused by a single point mutation in the sequence of a gene. For example, a change of one guanine base to cytosine in the *ras* oncogene, located on human chromosome 11, is frequently associated with bladder cancer. This simple change results in the amino acid glycine in the *ras*-encoded (Ras) protein being replaced at one point with a valine. This dramatically changes the function of the G-protein encoded by the *ras* gene. Normally, the protein cycles from an inactive to active state but the mutation does not allow this to happen and the protein is continuously active. Because the signal delivered by the Ras oncoprotein is continuous, the cell continues to grow and divide. This unabated growth leads to the bladder cancer (Bos, 1989).

Many drugs used in cancer therapy destroy or inhibit the growth of rapidly dividing cells and act at specific points in the cell division cycle such as the S phase. Unfortunately, they also affect non-cancerous rapidly dividing cells such as those in bone marrow, hair follicles and gastric mucosa. But the circadian variation in the timing of the cell cycle in these non-cancerous cells is often different from that in the cancerous targets. Therefore, if most of the daily treatment is confined to times of lowest S-phase or DNA-synthesis activity in the non-cancerous cells, the toxicity is reduced and consequently higher doses can be given.

In general, patients receive anti-cancer drugs at times that are convenient to the staff administering them. A growing body of data suggests that therapeutic effect may be maximised and toxicity may be minimised if drugs are administered at carefully selected times of the day. Dr Francis Lévi, at the Hopital Paul Brousse near Paris, and his associates in other European hospitals, have treated over 1,500 colon-cancer patients by infusing them with a rhythmically oscillating level of medication. The precise schedule depends on the drug. Lévi has found that the dosage of chemotherapy could be much higher than the maximum that the body can tolerate under a flat-rate regime. There was about a three-fold increase in the proportion of tumours that shrank by half or more. 'All the side effects are reduced by chronotherapy', claimed Lévi. 'And this despite the fact that the dose which could be delivered was higher by 40%' (Lévi et al., 2001).

The process of cancerisation has generally been considered to be more or less irreversible. There is growing evidence that this somewhat static linear biological view does not tell the whole or most accurate story of cancer development. The balance between cancer progression and health is dynamic.

Breast cancer, a disease that will be diagnosed in one of every nine living American women, is an example of the seasonal balance between a person and cancer. Several large studies, on each side of the equator, have demonstrated that the likelihood of the discovery of breast cancer is highest in spring, lowest in autumn and intermediate in winter and summer. These large studies were with culturally diverse populations living in very different climatic and geographic conditions. Since the average breast cancer takes many years to develop, these studies tell us that breast cancer size (which reflects both the growth of the tumour cells and the woman's defences against that growth) is growing more rapidly in springtime and less rapidly in other seasons. Other measures of breast cancer biology, including the average size of resected tumours, the pathologist's assessment of their microscopic aggressiveness, the number of lymph glands into which the cancer has spread, the concentration of hormone receptor molecules within the resected breast cancer cells and the overall survival of women with breast cancer, are each affected by season (Hrushesky, 1991).

Breast cancer is not the only cancer with yearly rhythmic (non-linear) biology. All of the pre-malignant and malignant changes of the surface of the uterine cervix are also non-randomly distributed around the year in a large population screened with microscopic examination of scrapings of the surface of the cervix, the so-called 'pap smears'. All pre-malignant and malignant changes of the cervical surface cells are more frequent in winter months.

Seasonal cancer biology is not limited to women. The discovery of testicular cancer, the most common cancer in men under 35 years of age, is also prominently seasonal. Cancers of the two common cell types of the testicle have almost opposite seasonality: the frequency of one type peaks in December/January, and the frequency of cancers of the other cell type peaks in August/September.

The realisation that the biology of the balance between a person and an incipient cancer on average is predictably different at different seasons for different types of cancer is useful to improve our understanding of cancer biology. This knowledge could increase our ability to prevent, diagnose and treat a cancer by optimally timing our preventive, investigative or therapeutic programme.

Preventive strategies may be most effective and most important to employ at certain times of the year. Mass screening programmes instituted at great cost for early cancer detection performed at the optimal season may generate fewer false negative, fewer false positive and more true positive test results, enhancing the effectiveness of cancer screening and substantially diminishing its costs. Therapeutically, careful seasonal modulation of treatment regimens may be more effective and less toxic. It has been demonstrated, for example, that chemotherapy is most damaging to normal tissue at certain seasons and that hormonal therapy may be more likely to be effective in some seasons (W. J. Hrushesky, personal communication).

Although chronotherapy requires an understanding of the full range of biological rhythms, most attention is on the daily circadian cycle. Heart rate varies over the 24-hour period, as does kidney function. The liver metabolises sugars, lipids and proteins at different rates over the course of the day. Recently, Charles Weitz and his team, of Harvard Medical School, compared more than 12,000 genes (over a third of the mouse genome) active in liver and heart during two days while mice were exposed to constant light. They found that between 8 and 10 per cent of genes in each tissue varied their activity following the 24-hour cycle (Storch et al., 2002). As most of the rhythmic genes are not the same in these tissues, the overall proportion of rhythmically expressed genes in the genome is probably higher than 10 per cent.

These genes are most likely to be involved in peripheral oscillators in these organs, and intimately coupled to the timing signal of the master clock in the SCN. Finding them suggests that there are probably oscillators in most tissues and that there is a considerable local time element in rodents and probably in humans, as well as what we might dub a Standard Mean Body Time. The linkage between these oscillators and the central role of timing in animal and plant physiology foreshadows the prospects of greatly improved therapies once the implications are fully understood.

For instance, impaired sleep and nocturnal restlessness are common in the 25 million people world-wide afflicted with Alzheimer's disease. People with dementia often develop an extremely disturbed sleep pattern. Disrupted and unpredictable sleeping patterns are frequently an important factor in the decision to admit a person with dementia to a home.

Traditional sedatives have limited usefulness and are accompanied by side effects. The poor sleep patterns are linked to other disturbances in the circadian system such as temperature and alertness. Eus Van Someren, of the Netherlands Institute for Brain Research, has pointed out that the problems are compounded because 'Alzheimer's patients often receive very low light and receive little physical activity. Both of these are very important to our biological clocks' (Van Someren *et al.*, 1999).

Bright-light therapy has been tried as way of helping the sleep/wake cycles of those with Alzheimer's disease as well as administering melatonin. In an attempt to improve the therapeutic interventions, David Harper, of Harvard's McLean Hospital, and his colleagues monitored the rise and fall of core body temperature and the waxing and waning of spontaneous motor activity in 38 dementia patients with either Alzheimer's disease or another form of dementia, frontotemporal degeneration (Harper *et al.*, 2001).

Body temperature reached its peak much later in the day in the Alzheimer patients than it did in controls. People with frontotemporal degeneration exhibited a different pattern. Their activity levels reached a peak earlier in the day than in controls, whereas their body temperature rhythms appeared normal.

Harper believes that the mixed success of the bright-light and melatonin interventions may be due to doctors not identifying the type of dementia that their patients suffer from. Chronobiological approaches may be more helpful to Alzheimer patients than those with fronto-temporal degeneration.

Light-exposure treatment may not have worked consistently in Alzheimer patients because the patients were exposed to bright light in the morning, often around 9.00 a.m. Yet in some patients, the circadian phase delay may be so great that it pushes the beginning of the circadian cycle (CT 0) – which usually occurs between 4.00 and 5.00 a.m. – past 9.00 a.m. In such extreme cases, the effect of light therapy would be to delay the cycle even more rather than advance it to a more normal position.

Harper and his colleagues observed that though the circadian rhythm was delayed in all Alzheimer patients, it was delayed by different amounts in different patients, a finding that clinicians could use to tailor more effective chronobiological treatments for their patients. 'When giving light or melatonin treatment, the most important thing to remember is that these patients have a subjective internal time,' Harper said. 'You need to measure the actual circadian rhythm of the patient before you administer light treatment.'

In ageing rats it is suggested that the SCN clock ticks over properly but the signals are not getting through. If there are problems with the output of the clock with age it may be possible to work on ways of improving the communication of the SCN with the rest of the brain to help in normalising sleeping patterns.

Rheumatoid arthritis and osteoarthritis are two other diseases whose prevalence increases with age and which have profound circadian rhythms in the manifestation and intensity of symptoms (Harkness *et al.*, 1982). Rheumatoid arthritis can be distinguished from osteoarthritis by the time of day when the patient's joints are most painful: morning stiffness is characteristic of rheumatoid arthritis, whereas symptoms often are worse in the afternoon and evening with osteoarthritis. Consequently, cyclo-oxygenase-2 inhibitors effectively relieve osteoarthritis symptoms when taken in the morning, but better results are obtained in rheumatoid arthritis when part of the dose is taken in the evening.

The general point is that timing is vital. Yet such a seemingly commonsensical and therapeutically beneficial procedure is still outside mainstream medicine. The value of chronotherapy was driven for many years by Franz Halberg and his colleagues at the University of Minneapolis. Halberg was insisting in 1975 (Halberg, 1976) that

At this time when a major effort of the US investment into medicine is said to be focused on the cancer field, chronobiologists have lacked salesmen good enough to persuade administrators to initiate a rigorous, properly scaled endeavour in clinical tests in chronotherapy. Chronobiology encountered benign neglect as national programmes focused on the search for new molecules. Almost certainly less than one-thousandth of the available National Cancer Institute budget is now used to exploit the already demonstrated possibility of improving treatment with new or old molecules according to circadian rhythms.

Cost is an issue and there are messy practicalities in the complicated, time-sensitive administration of the powerful, toxic drugs used in chemotherapy that do not fit well with the shift patterns and work schedules of busy hospitals. But second-generation, multichannel ambulatory pumps, with simplified programmability and reduced volume, weight and cost, should allow chronotherapy to be given to any patient as a cost-effective therapy – often in their own home. Although the issue of cost effectiveness needs further prospective assessment, there is a strong case being developed for the clinical relevance of chronotherapy and its integration into the early stages of anti-cancer drug development.

But there is still some way to go. Some doctors are simply not convinced of its value in their field of treatment. Although they acknowledge that chronotherapy obviously has some benefits, these are felt to be too small to warrant the use of the technique routinely. Karol Sikora, a former director of the World Health Organization cancer programme, has commented (Kmietowicz, 1997):

Chronotherapy is expensive, complicated to give, and very labour intensive. If we are devoting resources to this then they are being taken away from other parts of our work with cancer patients. It may be worth while using chronotherapy if we could identify which patients could benefit the most in terms of an improved response.

Biologists like Martin Ralph at the University of Toronto are sceptical, too (Aguilar-Roblero *et al.*, 2001):

Chrono-issues have the problem that they are important in the long term, but we can ignore them for short-term gain. So, any type of medical intervention that is based on this precept will take a back seat to an immediate cure. We can ignore short-term sleep deprivation if it will gain us an edge in business, for example, not considering the fact that we may impose the deprivation for years.

And Timothy Monk, at the same electronic conference organised by Sociedade Brasileira de Neurociências e Comportamento (SBNeC) in 2000, went even further: 'I find it very interesting that after all these years we do not have a real meaningful chronobiotic (i.e. a pharmacological *zeitgeber*). I wonder whether we ever will!'

Another reason for the low uptake of chronotherapy among doctors may be simple ignorance. Doctors are not entirely to blame – after all, it is not as though the procedure is jumping out at them from the literature. In August 2003, chronotherapy registered no hits when entered into the search box of the web sites of either the British Medical Association or the Royal College of Physicians. The *British Medical Journal* web site came up with 62 references. It is not just a British thing. The *New England Journal of Medicine* archive manages just 60 papers.

But chronotherapy is hardly brand spanking new, so there is no excuse of novelty: 25 years ago, in 1976, Georges E. Rivard of Sainte-Justine Hospital in Montreal began treating about 120 children with leukaemia with a combination of several anti-cancer drugs, including one known as 6-MP (for 6-mercaptopurine) that was to be taken at home once a day. He asked parents to give the drug to their children in either the morning or the evening, whichever was more convenient, but to stick to the same time throughout the treatment.

When Rivard analysed survival data five years later, he realised that the children who had taken 6-MP at night were three times more likely to have had their cancer go into remission. The effect has been lasting: Rivard published a follow-up study in 1993 showing that the children who had received chemotherapy at night were more than twice as likely to still be cancer-free. 'Almost everyone gives 6-MP at night because of this study,' he says (Rivard *et al.*, 1993).

The combination of drugs in the Rivard trial is no longer used. But the implication – that the timing of chemotherapy can increase its success – has been confirmed in both animal and human studies. Gaston Labrecque, a pharmocologist at Laval University in Quebec, points out (personal communication), 'We've been working in this field for 25 years, so we've known for about 25 years that the effectiveness and toxicity of drugs are not constant over a 24-hour period. But it takes about 20 years before it gets to be known by everybody doing research and another five to ten years before it's known to physicians in practice.'

Perhaps this is true, but it would be unwise to hold one's breath. The American Medical Association commissioned two extensive surveys on chronotherapy among doctors and the public back in 1996 whose results did nothing to support Labrecque's recent optimism: 320 primary-care physicians were surveyed by telephone, and a parallel survey was conducted with 1,011 adult members of the general public (Smolensky, 1998).

When the doctors were asked how familiar they were with chronobiology, described as 'the study of the body's circadian rhythms and their effect on human biological processes', only 5 per cent said they were very familiar with the concept, 44 per cent said they were somewhat familiar, while 39 per cent were not very familiar and 12 per cent professed not to be at all familiar.

Even a generous interpretation of these findings would have to suggest that the majority of American doctors had little knowledge of the concept, let alone the practice, of chronobiology. In reality, the percentage who had any real understanding was probably around one in five, if that.

This lack of knowledge was evident when the doctors were asked about the temporal rhythms of a range of medical conditions. The selected conditions were: angina pectoris and myocardial infarction, to which vulnerability is highest in diurnally active at-risk persons between 6.00 a.m. and noon; the rapid morning rise in blood pressure; the morning intensity of arthritic pain; the night-time peaking of asthma; and the heightening of symptoms of migraine headache and allergic rhinitis between 6.00 a.m. and noon.

These are common conditions and yet, as Table 13.1 shows, in some instances only a quarter of the doctors sampled knew the correct timing of the event (Smolensky, 1998).

If the doctors' knowledge of chronotherapy was poor, unsurprisingly that of the public who were surveyed was abysmal. But, bad as it was, more of them than the doctors knew that the symptoms of allergic rhinitis – better known as hay fever – are worse in the morning.

Perhaps the most startling finding was that only 26 per cent of doctors seemed to know that asthma attacks are most prevalent at night. Asthma attacks are several hundred-fold more likely at night than dur-

Table 13.1. US doctors' knowledge of temporal rhythms of medical conditions

Medical condition	Percentage of doctors knowing correct time of event
Blood pressure	45
Cardiovascular system	40
Angina pectoris	38
Asthma	26
Migraine headache	24
Rheumatoid arthritis	46
Allergic rhinitis	24

ing the day. There is a marked diurnal rhythm underlying asthma that shows itself by an increase in bronchial hyper-reactivity, disturbance of sleep, variable lung function at night and early-morning tightness of the chest and wheezing.

Asthma has been known for an awfully long time. It gets its name from a Greek word meaning to breathe hard or to pant and was described by Hippocrates in the fourth century BC. But it had been known to the Chinese at least 1,500 years earlier. They treated it with Ma Huang, the leaves and stems of the *Ephedra* plant, and 4,000 years later ephedrine, found in species of *Ephedra*, was used as a treatment in the West. Maimonides, the great philosopher-doctor, described it in detail in the twelfth century and specifically noted its night-time incidence. 'The night belongs to asthma,' says Richard Martin, MD, Head of the Division of Pulmonary Medicine, National Jewish Centre for Immunology and Respiratory Medicine. 'Morbidity and mortality is greatest during sleep-related hours ... if you control the night-time asthma, you can change the severity of asthma' (personal communication).

Since the greatest prevalence of asthmatic symptoms occurs around 4.00 a.m., Professor Martin reasons that medication therapies should be timed to take effect when inflammation in the airways is the greatest. However, he adds, 'most physicians are taught – it doesn't matter if it's asthma or heart disease – divide medications equally throughout the day.' Even though the diurnal nature of asthma was recorded nearly 1,000 ago – long enough that one would have expected it to get onto the curriculum of even the most hard-pressed medical school – the lesson has still not been learned by about three-quarters of US doctors.

In fairness, things have improved a little. In 2002, the Gallup Organisation interviewed 200 American primary-care physicians in a survey sponsored by the Biovail Corporation. Two out of three physicians (66 per cent) reported that they were familiar or very familiar with the concept of circadian rhythms, and only 1 per cent admitted total ignorance. But of the 99 per cent who had some familiarity, only one in six (17 per cent) believed that knowledge of the body's circadian rhythm is very important to the treatment of disease (Biovail Corporation, personal communication).

The American experience is almost certainly typical of other countries. The obvious question is why this should be if chronotherapy offers substantive advantages. Why has it not permeated through to medicine and pharmacy? Medical students are still taught the dominant paradigm that most bodily processes are in a state of homeostatic equilibrium and do not change significantly over time. The homeostasis paradigm implies that the probability of risk or intensity of disease is equal throughout a specific period, even though this is seldom the case. Franz Halberg's mantra, that predictable variability should be regarded as a friend and not as a foe, has made little headway.

But that raises another question. Why is it that medicine has not looked beyond the homeostatic paradigm? One good reason could be that the evidence for the widespread effectiveness of chronotherapy is not yet in, despite its undoubted effectiveness in some cases. As Martin Ralph stated bluntly at the SBNeC, 'Intelligent chronotherapy will only enter the main stream if it can be shown to have unequivocal advantages.'

So there was considerable hope among the chronotherapy community for CONVINCE, a large clinical trial of Covera-HS, a hypertension drug marketed by Pharmacia. Cardiovascular disease has been the main cause of death in developed countries for over 100 years. In 1997, 31.4 per cent of Americans who died in that year died of heart disease, and 6.9 per cent died of cerebrovascular disease (Hoyert et al., 1999). One argument, among many, for its high incidence has been a lack of appreciation of the role of chronobiology. Cardiovascular death is 33 per cent more common during December and January than during the summer, even on the West Coast of the USA, where climatic variation is less extreme than in many other parts of the world. Heart attack is about 33 per cent more common among working people on Monday, compared with the rest of the days of the week. The greatest intrinsic variation in cardiovascular events, however, is associated with the circadian pattern, with a peak during the early morning (Kloner *et al.*, 1999).

The risk of heart attack from 6.00 a.m. to noon is some 30–40 per cent higher than would have been expected if heart attacks occurred randomly and evenly throughout the 24 hours. There is a similar heightening in risk from stroke between 6.00 a.m. and noon. The most likely reason for the rise in cardiac events is the temporal correlation of blood pressure, heart rate and cardiovascular risk, all of which peak in the 6.00 a.m. to noon period (Appendix I). Blood pressure is at its lowest during the night and rises about 10–25 per cent from 6.00 a.m. to noon. The pulse rate also has a night-time low and increases before awakening, again by about 10–20 per cent. The blood pressure surge places stress on the arterial walls and if there is an unstable coronary or carotid plaque present there is an increased probability of its being dislodged, then blocking a coronary artery and causing a heart attack. Blood pressures above 130/90 increase the risk of heart attacks and strokes by factors of up to fivefold.

If the most dangerous period is during the morning, it would seem sensible to ensure that any drugs being taken to reduce blood pressure are effective at that time. The hour of day at which you take a drug can greatly alter its effectiveness. Morning is the time for high-blood-pressure medication to be working at its peak. Night-time dosing allows blood pressure medication to reach peak concentrations as morning circadian variations begin to appear.

The Biovail/Gallup survey showed that although 55 per cent of the US primary-care physicians interviewed knew that blood pressure peaked in the morning, only one in seven (14 per cent) patients are advised to take their blood pressure medication at night or at bedtime. Stephen Glasser, professor of epidemiology at the University of Minnesota's School of Public Health, says (Glasser *et al.*, 2003):

Traditional antihypertensive medications are not designed – and are clearly not being administered – in a way that adequately addresses increased morning risks. To maximise the many advantages of once-daily dosing, physicians must ensure that peak concentrations are in effect when patients need them most. Given the prevalence of morning administration, most patients enter the 'am surge' with the lowest concentrations of the day.

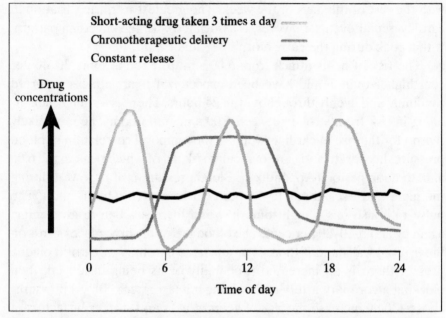

Figure 13.1. A representation of drug delivery using three different timing strategies. A circadian-efficient delivery is compared with standard formulations that may be taken two or three times a day and produce a 'spiky' drug level and a once-a-day controlled-release formulation that provides a 'steady' drug level.

Covera-HS was designed to be a once-a-day 'circadian efficient' drug for use in hypertension. The HS stands for *hora somni*, as it was to be taken at bedtime. It is based on verapamil, a well-known calcium channel blocker used in lowering blood pressure by relaxing the small blood vessels. The idea was to deliver a dynamic drug level that synchronises the medication to the blood pressure rhythms and so optimise the desired effects and preferably minimise any of the side effects (Figure 13.1).

The drug was delivered by the clever design of a capsule with a hard 'delay' coat that takes several hours to dissolve in the patient's gastrointestinal tract. When it dissolves, several small pellets, each about a millimetre in diameter, are released. Each pellet has an inert core surrounded by the active drug, verapamil, wrapped in a sandwich of polymer layers. As the soluble layers dissolve the active ingredient is slowly 'pushed' out.

It seemed like a lot of trouble to go to when the simpler answer might be to take ordinary hypertension tablets at night. But blood pressure is tricky. It falls at night anyway, and among older people in particular an even greater 'dip' can be damaging. Further, the reduction in blood pressure is different when you take medication in the morning than when you take it at night. The clinicians were keen on a better treatment and controlled-onset extended-release verapamil (Covera-HS) looked a good prospect. Simple trials showed that it was effective in lowering morning blood pressure, but the key question was not whether it lowered blood pressure, but whether it was more effective in preventing cardiovascular events than other treatments.

The only way to find out was a large, randomised trial that would run for long enough to pick up enough cardiovascular events to provide a statistically meaningful analysis These big trials are enormously difficult to set up and very expensive to run (Black *et al.*, 2001). However, 16,602 hypertensive patients over 55 years of age were recruited in over 600 clinical sites in 15 countries. They were randomly assigned to receive either the chronotherapeutic formulation of verapamil in the evening or more traditional therapy with atenolol or hydrochlorothiazide in the morning. The trial was scheduled to run for five years.

After three years Pharmacia pulled the plug. Henry Black, of Rush Presbyterian–St Luke's Medical Center and the organiser of the massive trial, explained that because of the premature termination the findings are inconclusive. Covera-HS was as good as the standard treatments in lowering blood pressure, but there was no clear evidence that it was any better at preventing heart attack, stroke and cardiovascular death (Black *et al.*, 2003). The editors of the *Journal of the American Medical Association*, in an accompanying editorial, described the premature ending of the trial as unethical.

Clinical trials are sometimes stopped prematurely. The drug being trialled may be seen to be harmful; the research may be proving futile; or the trial may be badly designed and administered. Sometimes, the new drug may be so effective that it is considered unethical not to put the patients in the control arm of the study onto it immediately. But occasionally a trial is stopped for commercial reasons that have nothing to do with the clinical investigation. Whatever Pharmacia's motives, Stuart Pocock, a professor of medical statistics at the London School of Hygiene and Tropical Medicine, called CONVINCE 'the single most important randomised trial of the moment entering the calcium-antagonist debate. It's a crying shame that the sponsor did not allow true resolution of the data' (Moore, 2002).

Professor William B. White, of the University of Connecticut School of Medicine, was another of the principals in the study. He accepts that 'chronotherapy took a major blow with CONVINCE'. But he still thinks (personal communication) that

> we have ample evidence that cardiovascular chronobiology exists and is relevant to scientists and clinicians involved in human research and medical practice. Despite the lack of impact in the CONVINCE trial, there still is an important role to study potential interventions that link effectiveness in reducing cardiovascular events to chronobiologic principles.

The CONVINCE trial on its own would not have confirmed the value of chronotherapy to the sceptics; and, likewise, its inconclusive ending has not condemned the application of chronotherapeutic principles to medicine to its adherents. But it is fair to say that its unfortunate early demise puts back the likelihood of Smolensky's hopes that

> To truly embrace chronomedicine, physicians will need to abandon many of their current practices. They will need to throw out many norms by which they now evaluate patients. They will need to write in their charts not only what signs and symptoms they find, but when they find them. They will need to pay attention to the time of day they draw blood, and when they collect urine and other bodily tissues for diagnostic tests. They may even need to schedule tests at specific times of day or night.

Despite some 40 years or more of evangelising and several lines of clear evidence as well as some inconclusive leads, awareness of chronotherapy is still low in the medical and pharmaceutical world. In answer to the question why, Richard Martin replied (personal communication):

> I would like to give an evidence-based answer to the question, but there is none. If any subject is not taught in medical school and not stressed in training, then it is very difficult to have physicians change the way they practice. Since this, to my knowledge, is not done, it would take a

concerted effort by an institution such as the NHLBI [National Heart, Lung and Blood Institute] of the National Institutes of Health in the USA to put forth guidelines for this topic. This will never happen. Thus, it will be a very slow process backed by hundreds of research publications and editorials.

But the clock, so to speak, is ticking. A team led by Chang Chi Lee at Baylor College in Texas has shown that not only the *Per2* gene plays a key role in controlling the circadian rhythm in mice but that mice deficient in the *mPer2* gene are cancer prone. Lee and his colleagues suggest that the gene is important in regulating DNA-damage-responsive pathways (Fu *et al.*, 2002).

Linking clock genes directly to tumour growth ups the ante with regard to the role of circadian timing in disease, and although Richard Martin's pessimism is justified by experience, it could be that at last the importance of the biological clock will be appreciated in the wider medical community.

14

FUTURE TIMES: UCHRONIA OR DYSCHRONIA

Homo sapiens, the first truly free species, is about to decommission natural selection, the force that made us ... Soon we must look deep within ourselves and decide what we wish to become.

EDWARD O. WILSON (WILSON, 1998)

The US Defense Advanced Research Projects Agency is searching for ways to create what it calls the 'metabolically dominant soldier'. Among the projects it is pursuing in the Continuous Assisted Performance programme is the creation of a warrior who can fight 24 hours a day, seven days straight. 'Eliminating the need for sleep while maintaining the high level of both cognitive and physical performance of the individual will create a fundamental change in war-fighting and force employment,' says the Defense Sciences Office (Groopman, 2001).

Indeed it will. Various researchers have suggested that one night's loss of sleep results in a 30 per cent loss in cognitive performance, rising to 60 per cent after two nights (Buguet *et al.*, 2003). Soldiers, sailors and aircrew have to make instant decisions based on incomplete information, so a 60 per cent drop in cognitive performance really can be the difference between life and death.

British troops used stimulants to keep them awake them during the Falklands conflict, and USAF aircrews took amphetamines during the Libyan air strikes. But aside from their addictive properties, ampheta-

232

mines cause a range of side effects from agitation, irritability and nausea to increased heart rate, tics and impotence. When the drug wears off, it can lead to a rebound effect that causes extreme fatigue or depression

More recently the French government admitted that its crack corps, the Foreign Legion, used modafinil during covert operations inside Iraq during the first Gulf war. There are great hopes for modafinil among the US military, which is allegedly spending $100 million on research on the rationale that soldiers who sleep less will give the USA a military edge. Although nobody will admit it officially, it is likely that modafinil was used in the second Iraq war (Morris, 2003). Professor Michel Jouvet, an authority on sleep, claimed during an international defence meeting in Paris that 'modafinil could keep an army on its feet and fighting for three days and nights with no major side effects' (Morris, 2003). It has been used in some sections of the Belgian, Dutch and US airforces.

Modafinil is a so-called eugeroic ('good arousal') drug. In 1998 it was given FDA approval in the USA for use as a treatment for daytime sleepiness in narcolepsy. Although its mode of action is still being studied, it seems that it stimulates orexin-producing neurones in the hypothalamus. Orexins are a recently discovered family of neurotransmitters that are involved in the regulation of the sleep/wake switch.

But there is not much money in making drugs for narcoleptics, and modafinil is now marketed as Provigil in the USA for 'shift-work sleep disorder', excessive sleepiness caused by odd working hours, or more probably through shift-work patterns that demand phase shifts from an innate circadian rhythmicity that cannot respond that quickly. This is compounded by living in a visual environment that is created by electric light. Electric lighting does not provide ideal intensities, spectra and timings for the circadian system. Modern lighting practice has effectively led to a disassociation of light for vision, which is adequate, from light for the circadian system, which may be inadequate.

Police, hospital staff, pilots and people who work in all-night stores are among millions of workers in our 24-hour society who are likely to be affected. As one TV programme put it, 'Imagine a pill that would make sleep unnecessary for fighter pilots on long-range missions, or even the high-powered executives and parents of newborns among us' (ABC, 2001).

Why stop at drugs? Future warriors will face intense, around-the-clock fighting for weeks at a time. A genetically cloned, fully awake, fully functional '24/7' soldier, sailor or aircrew who needs no sleep but remains effective and happy would be even better. As the 24-hour society intensifies, will bioengineering eventually produce citizens with a variant of Morvan's syndrome* without the deleterious effects, who will be biologically adapted to work nights?

It is not far-fetched to imagine that in the next few years we will learn how to manipulate our circadian rhythms and so disconnect ourselves from the natural world. Welcome to a world in which we sleep two hours a night. In what some might see as yet another conquest in a battle to subjugate nature we could, metaphorically, stop the sun and so regulate our behaviour free from the celestial rhythms, as happened on the plains of Gibeon:

> *the sun stood still in the midst of the heaven, and hasted not to go down about a whole day.*

> JOSHUA 10.13

Circadian rhythms seem to be universally vital to life and may in fact be indicative of it. It has been suggested that a biosignature detected during the 1976 Viking mission to Mars could only have been made by life forms. In one experiment a radiolabelled (^{14}C) nutrient solution was added to a Martian soil sample and the subsequent evolution of radioactive gas was observed. The amount of gas released fluctuated, and a close examination of the old data by Joe Miller at the University of Southern California suggests that, at steady state, these fluctuations exhibit a periodicity of 24.66±0.27 hours, statistically indistinguishable from the Martian solar period. It takes some swallowing, and there are many doubters, but Miller considers that subsequent analyses of the entire Viking LR dataset support the conclusion (personal communication).

The rhythms regulate many of the most important and intimate aspects of behaviour and answer key biological questions. Many species of *Drosophila* eclose at dawn. Is this in response to an environmental stim-

---

* Morvan's fibrillary chorea or Morvan's syndrome is characterised by a range of symptoms including severe loss of sleep (agrypnia). Michel Jouvet and his colleagues in Lyon, studied a 27-year-old man with this disorder and found he had had virtually no sleep over a period of several months. During that time he did not feel sleepy or tired and did not show any disorders of mood, memory or anxiety. Nevertheless, nearly every night between 9.00 and 11.00 p.m., he experienced a 20–60-minute period of auditory, visual, olfactory and somaesthetic (sense of touch) hallucinations, as well as pain and vasoconstriction in his fingers and toes (Fischer-Perroudon *et al.*, 1974).

ulus (a change in light, temperature, humidity) or in response to an endogenous timing mechanism hard-wired into the fly? The answer, as we
have been at some pains to describe, is that it is endogenous. A bird may
fly in one direction for a few thousand kilometres and then turn to fly in
another direction. Is the bird just responding to landmarks or magnetic
cues, or does a biological clock determine how long the bird flies in a
particular direction during migration? The answer again is that it has an
internal clock that plays a vital role in migration.

In 1960 Colin Pittendrigh remarked, 'there are common mechanisms – built of different concrete parts – in circadian systems and
photoperiodic effects everywhere' (Pittendrigh, 1960). The past 40 or
more years have proved him right. In the process, chronobiology has
emerged as a scientific programme that cuts across physiology, biochemistry, neuroscience, evolutionary biology, behavioural genetics,
ergonomics, ethology and ecology, and wanders into the realms of
anthropology and philosophy.

Chronobiology may be defined today as a multidisciplinary effort to
understand the temporal dimension of life, being mostly recognised by
the study of the phenomena known as biological rhythms. Among these,
circadian rhythms and the endogenous mechanisms that generate sustained oscillations in all levels of biological systems have been the most
prominent. It has become an organising principle, producing advances
in the understanding of biology and medicine, and it has profound implications for the organisation of human activity.

Pittendrigh himself did not like the word chronobiology. Although it
has become the recognised term for many of those who study biological
rhythms, he regarded it as 'something I oppose because it is unnecessary, pretentious and inaccurate'. To his mind, a separate discipline to
study temporal phenomena such as rhythms 'would be comparable to
some physiologists like Bernard suggesting there ought to be a society
for the study of homeostasis' (Cambrosio & Keating, 1983). Pittendrigh
considered himself a circadian physiologist and an evolutionary biologist, someone with a special interest in the 24-hour rhythmic activity
characteristic of living organisms. He and his colleagues saw such studies in terms of a specialisation within biology. The research aim was to
demonstrate unequivocally that endogenous biological clocks existed, to
develop the idea of the analogy between the clock and a self-sustaining
physical oscillator familiar to physicists and then to move on to locate

the clock and its mechanism and entrainment by light. It is a reduction-ist programme that isolates the components of a system and then puts them back together again. Of course, there may well be properties that emerge only at the system level, as Pittendrigh himself thought, but it is a programme of classical cause and effect. Broadly speaking, it is his ap-proach that has been described in this book.

It has been a very successful programme. For biochemists and physiologists, circadian rhythms were useful tools as they were easy to measure, being precise, stable and usually refractory to disruption by extraneous variables. As we saw, Benzer and Konopka were able to use a *Drosophila* behaviour to screen mutants and so find the *per* gene. For students of behaviour as such, many behavioural circadian rhythms are easy to read automatically and they often have clear meaning in the lives of the animals that perform them. The circadian clock provides a unique means of investigating how a cellular phenomenon is adapted to the environment, ultimately resulting in specific behavioural patterns. Menaker's prediction in 1981 was proved right (Menaker & Binkley, 1981):

> *it seems possible that we will have an understanding of the mechanisms that generate and control circadian behaviour before such understanding is achieved for other behaviours of equal complexity and importance.*

But the story has not been a single track. Another group of what might be termed 'hard' chronobiologists formed around Franz Halberg in Minneapolis. Their agenda, which came from a medical rather than a biological standpoint, was wider from the start. They were interested not only in circadian rhythms but also in the problems of growth, develop-ment and ageing; 'in other words', according to Cambrosio and Keating, 'those concerned included not only biologists, but gerontologists, paedi-atricians and, in general, medical practitioners' (Cambrosio & Keating, 1983). Pittendrigh's oscillator model was regarded as more of a hin-drance than a help by Halberg, 'as an obstacle which must be overcome in order to begin the modern study of rhythms based on the use of infer-ential statistics' (Cambrosio & Keating, 1983). Crudely put, Halberg and his colleagues were mostly interested in the hands of the clock; what were the outputs of rhythmic activity and how could they be manipu-lated to benefit medical practice.

Relationships between the two schools were not always friendly – 'clockwatchers' was the contemptuous term given to those trying to unravel the mechanism of the clock. Pittendrigh and his friend Aschoff rejected the idea of a separate discipline, as they believed in gathering an ensemble of scientists from different backgrounds to engage on a particular biological problem. Halberg saw the study as a discipline in its own right with its own practices, tradition and even a constitution.

Whatever the difficulty of the antecedents – and echoes of the past still hang around today, as science is still very much a social act by flawed human beings – the past few decades have been incredibly productive, helped by two factors, one financial and one technical. The financial impetus came in two waves. First, the programme of manned space flight in the 1960s brought in money for those scientists studying the possible impact on the astronauts of the rapid changes from day to night as the spacecraft orbited the earth. The second large wave of money came when the National Institutes of Health named the 1990s the Decade of the Brain and money flowed into neuroscience, including the study of rhythmic behaviour.

Technically, chronobiology has grown up in the period when genetics has been the dominant organising principle in biology. It is 50 years since Crick and Watson described the structure of DNA, but it was their laconic understatement 'It has not escaped our notice that the specific pairing we have postulated immediately suggests a possible copying mechanism for the genetic material' that kick-started a genetics triumphalism that has rolled over everything before it, culminating in the draft of the human genome.

The chronobiologists swam with the tide and, using circadian rhythmicity as the behavioural marker, they uncovered a series of 'clock' genes. Pittendrigh's hope that the mechanism of the clock would be gradually uncovered, and with it an understanding of how the rhythms worked, was on target. But the 'one gene – one protein' mindset that has dominated biological thinking is meeting challenges. Sceptics raise simple queries. For instance, how is it, they say, that we can claim that there are 30,000 genes in the human genome?

It is an interesting question. After all, several gene products can be made from one gene by cutting and splicing the primary transcript in alternative ways. And what is included in a gene? Do we include the promoter and regulatory sequences that affect whether the gene will be transcribed – no easy task when regulatory regions can be distant from the rest of the gene and can be involved in the regulation of more than one gene. Actual coding regions of DNA may also form part of two over-lapping genes. And the sharing of regions of DNA by different genes poses a problem for the idea that a structural definition of a gene can be given independently of information about the role that a DNA region has in some larger system (Moss, 2003).

Lenny Moss, a member of the philosophy department at the University of Notre Dame, Indiana, has pointed out (Moss, 2003):

no one yet even claims to know the phenotypic function of every gene in a genome, it would appear that such knowledge is not necessary for individuating and tallying up genes on the basis of molecular sequence data. So how is it done? What are the criteria that enable a genome project to tell us how many genes humans, flies, worms, yeast or rice are endowed with?

The answer, according to Moss, is:

investigators have abstracted from the features of particularly well-characterised molecular genes a set of criteria none of which are necessary and none of which are sufficient, and which needn't necessarily occur together, and which by no means are exhaustive. Two of the most salient examples of such criteria would be evidence that the stretch of DNA in question serves as a template for the synthesis of an RNA transcript and the presence of an AT (adenosine-thymine) rich 'TATA box' promoter site.

When we say there are 30,000 genes in the human genome we are making an arbitrary statement and in so doing we are putting forward a specific mental model as to what we mean by genes and the way they work. The route from an identified gene to a functional protein is not straightforward. A gene may be switched on, but its messenger RNA need not be translated into protein. Genes often produce more than one type of protein, for example by alternative splicing of messenger RNA, and these proteins may have different functions. Proteins are commonly modified after they have been constructed; pieces of the protein may be cleaved off, or other molecules such as lipids and sugars can be added. With all these possibilities, it has been estimated that the human proteome (all proteins manufactured) is at least an order of magnitude more complex than the human genome.

These big epistemological problems in biology have considerable consequences. Although mutations undoubtedly play a part in many cancers, the full story is far more complex. Moss has two explanations for the ever-increasing emphasis on genetic screening. One is the story of the drunk looking for his keys under the lamppost: 'That's where the light is'. Though genes may be an important part in the origin of only a small number of tumours, and almost never sufficient by themselves, at least we know how to look for them. Second, genetic screening can be marketed to patients. Another description by Olden and Wilson has it that 'The blueprint for studying cancer in relationship to the environment is changing from reductionist models, which look at one or two variables and a few genes, toward a complexity theory, which examines multiple systems' interactions' (Olden & Wilson, 2000).

Craig Venter, one of the architects of determining the draft human genome put it all more colourfully: 'We don't know shit about biology' (Preston, 2000).

Because the circadian rhythm is the best-characterised gene–protein–behaviour mechanism in biology, it has considerable heuristic significance in developing new and more general models of biological understanding. Till Roenneberg's idea of a circadian network composed of a complex brew of interacting genes and proteins that in turn may individually also belong to other networks produces a richer, more complex mix than the single central loop made up of dedicated genes monotonically beating out a rhythm.

A self-correcting, self-organising circadian network tuned to a daily 24-hour rhythm is a different metaphor from the mechanical clock which dominated earlier circadian thinking. The mechanical clock was the symbolic machine of the Industrial Revolution. It was supremely rational. An intelligent being could take it apart and put it back together again and it would still work. It was a mechanism of the type that Immanuel Kant described as having a functional unity in which the parts exist for one another in the performance of a particular function. Pre-existing parts were machined and assembled and the clock ticked. And once assembled it was not only the controller but also the coordinator. Whosoever controlled the clock controlled the work rate. Workers who retire are given a clock in symbolic memory of the time when the mill owners forbade their workers to bring in their own clock to the factory. It was not unknown for the self-same mill owners to adjust the clock they controlled so that it ran more slowly during the day and consequently the workers did more than their paid share.

Circadian rhythms are about timing rather than time itself, defined as regulating actions to produce the best effect. The player swings the tennis racquet so that it connects with the ball at precisely the right instant for maximal effect, and the commentator drools about perfect timing. We talk about dramatic timing; comic timing; timing for best effect. Timing is of the moment. It is about time to and time of. This is time as opportunity in the language of Ecclesiastes 3. 6–8:

> *A time to get and a time to lose; a time to keep and a time to cast away;*
> *A time to rend and a time to sew; a time to keep silence, and a time to*
> *speak;*
> *A time to love and a time to hate; a time of war and a time of peace.*

Our internal clocks enable us and the rest of the natural world to optimise time as moment. Time is a slippery concept. We say 'about time' when we admonish someone for being late; we ask what is the time when we mean clock time, but then ask, 'Are you having a good time?' which is an inquiry about an emotional state. Back in the fifth century it caused St Augustine no end of trouble when he reflected on the meaning of time. He concluded, 'If no one asks me I know, if I wish to explain to one that asketh I know not' (Augustine of Hippo, 1998). A millennium and a bit later, the realisation that Newton's Laws of Motion worked equally well if time runs forward or backwards put the proverbial cat among the pigeons. The cosmologists and the philosophers have had a field day with time. In a nutshell, is time moving forwards and, if so, through what, or are we moving forward through time? Does time exist at the most fundamental level? Solve that and it is next stop Stockholm and the Nobel Prize.

It is our relationship with the notion of time that makes us human. Probably alone among living creatures, we are aware reflexively that there are consequences to our behaviour that will affect not only ourselves but also others. By definition, consequences happen in the future, they are time dependent. We are time-knowing animals, and to deal with these consequences that result from our freedom of action we have de-

veloped codes and laws to regulate our behaviour. We have moved beyond behaviour to conduct.

Time-knowing enabled our ancestors to become farmers. They had
to anticipate the seasons. Birds do this automatically through the innate,
hard-wired photoperiodic signal. We are not photoperiodic. We had to
learn about the rhythm of the seasons as Aristophanes described in *The
Birds* (Aristophanes, 1978):

> *All lessons of primarily daily concern*
> *You have learned from the Birds, and continue to learn.*
> *They give you warning of the seasons returning.*
> *When the cranes are arranged and muster afloat*
> *In the middle air, with a creaking note,*
> *Steering away to the Libyan sands,*
> *Then prudent farmers sow their lands.*
> *The shepherd is warned by the kite reappearing,*
> *To muster his flock and be ready for shearing.*
> *You strip your old cloak at the summer's behest*
> *In assurance of summer and purchase a vest.*

Birds do not have to learn about time. They do not fret, unlike St
Augustine, over the question, 'If the past is over, and the future has not
yet come, all that exists is now; so how long does now last?'

We have a temporal order through our circadian clockwork, as do
other living creatures, but we go further, we can construct and feel time.
When we speak of time flying or dragging we are describing a relativity
that to us is both real and perceived. Alcohol or nitrous oxide seem to
reduce 'felt' time compared with 'clock' time, in contrast to cannabis,
which prolongs it. It has been argued that 'disinhibition' in the brain
allows a greater than normal sensory input, and this could give rise to
more numerous mental impressions than the usual number of mental
impressions per unit time, and so the greater 'felt' time. Time seems to
pass slowly when we don't do much. How we perceive time has dogged
psychology since William James wrote about it in 1880. It is unlikely to
be solved in this century.

But we do understand, in considerable molecular detail, how a
rhythm generated by the synthesis and degradation of proteins is
entrained by light into a cycle that is close to that of the solar period and

how this rhythmic information is transmitted to other biochemical processes in the organism. To answer the question in the introduction to this book, we are beginning to understand how our internal rhythms are involved in determining why we eat when we do, drink when we do, sleep when we do, and so on.

We need to. Since the first electric lights were installed in domestic homes in the 1880s we have markedly changed the natural photic environment and, in E. O. Wilson's words, 'decommissioned natural selection'.

Over 20 per cent of the working population now work at least some of the time outside the normal 7.00 a.m. to 7.00 p.m. working day, and the trend is increasing. Living outside the normal circadian pattern has undoubted health risks. But we have choices. We can use what we know about the molecular mechanisms of circadian rhythms to mitigate the biological harm of our '24/7' world. This may be through drugs, light, or manipulation of the workplace environment.

The BBC News Service on the Internet carries the slogan 'updated every minute of every day'. Human life has been shredded into small fractions of linear time. Our organisational and institutional thinking now favours what Nowotny (1994) calls:

> the scheduled and restricted. Managers who work in organisations which set deadlines for them, and who are expected to set deadlines for others, attempt for their part to cope with the chronic lack of time. Every hour in the appointments diary is organised by secretaries or assistants, whose task it is to negotiate. To buy and sell their boss's time.

Can we create a world in which we can manipulate time to offer a time paradise or 'Uchronia' for a time-stressed populace? Or will it be the time hell of 'Dyschronia'?

Again, we must distinguish between timing and time. There is evidence that circadian rhythms are involved in a wide range of psychological disorders such as depression. It is becoming clear that light has a key role in the sleep behaviour of older people, and the lack of it may well be instrumental in the disturbed sleep of Alzheimer patients. Despite the difficulties, there is still a considerable belief that much of general medical practice would be improved by taking a chronotherapeutic approach.

But we will be able to go much further. We will probably be able to gain time by reducing the amount we need to sleep. We may learn to change the perception of time by altering our circadian rhythmic states so that we slow time down. Herodotus tells the story of the Egyptian king for whom an oracle had predicted death in six years: 'perceiving his doom was fixed, he had lamps lighted every day at eventime … and enjoyed himself turning nights into days and so living twelve years in the space of six' (Herodotus, 2003). Our mass-market version of burning the candle at both ends may be as simple as taking a pill.

Or we could change the way we live. The late Lord Young acknowledged the tendency to change the environment rather than ourselves when he noted, 'People will not wear more clothes in the winter to keep warm if they have the option of heating much larger spaces in which to sit with almost nothing on' (Young, 1998).

How we think and feel about time largely determines our view of the world. Native American cultures have always believed that we had to be in synchrony with nature. Great care had to be taken as to how an individual or a group addressed the issues of nature. It is a different way of seeing the world – one group rushing away from nature, the other group rushing to a place, one group seeing linear time, the other group seeing circular time.

John Mohawk explained in a BBC talk (Mohawk, 2000):

> *The Iroquois, for example, thought that what we do today will reverberate for a long time in the future so there was a true conservatism, not conservative in the right/left politics but conservative in the sense you would be very careful about what it is that you did now because you would understand that it would have magnified effects down the road. There was that idea that people had to act today in responsibility for things that would happen seven generations into the future.*

We now live in a world in which we complain if the doors on the lift take too long to close. We are poised to alter our sense of time. We need to look deep within ourselves and decide whether we are sure we are going to become time-wise and not time-foolish.

GLOSSARY OF COMMON TERMS

Some terms in this glossary are from J. Aschoff, *Biological Rhythms Handbook of Behavioral Neurobiology*, Volume 4), Plenum Press, 1981 and from R. Reffinetti, *Circadian Physiology*, CRC Press, 2000.

Amplitude. (1) Difference between maximum (or minimum) and mean value in a sinusoidal oscillation. (2) Difference between maximum and minimum value of a biological oscillation.

Arabidopsis. *Arabidopsis thaliana* (thale cress). A plant used widely in genetics research.

Biological clocks. Self-sustained oscillators which generate biologic rhythms in the absence of external periodic input (for example, at the gene level in individual cells).

Chromophore. The light-absorbing molecule in a photopigment complex. The chromophore of animal photopigments is a specific form of vitamin A (11-*cis* retinaldehyde) that is bound to a protein called an opsin.

Chronobiology. Derived from the Greek (*chronos* for time, *bios* for life, and *logos* for study), the word is used to denote the study of biological rhythms.

Chronopharmocology. The practice of administering medicinal drugs in time schedules that optimise the therapeutic action of the drugs (in consonance with the patient's circadian rhythms).

Chronotherapy. Use of treatment timed according to the stages in the sensitivity-resistance cycles of target (or non-target) tissues and organs (or of the organism as a whole) to enhance the desired pharmacological effect and/or to reduce undesirable side effects of drugs or other therapeutic agents.

Circadian rhythm. A biological rhythm that persists under constant conditions with a period length of around a day. From the Latin *circa* and *diem*, 'about a day'.

Circadian time (CT). The subjective internal time of an organism under constant conditions. By convention, CT 12 corresponds to activity onset for a nocturnal species, whereas CT 0 designates activity onset for a diurnal species.

Circannual. A rhythm with a period of about 1 year (± 2 months), synchronised with or desynchronised from the calendar year.

Clock gene. Gene involved as a component of the molecular mechanism that produces a circadian oscillation.

Clock-controlled gene (CCG). A gene whose expression is regulated directly by the core oscillator mechanism.

Cyanobacteria. Photosynthetic bacteria, sometimes called 'blue-green algae'.

Dampened oscillation. Oscillation decreasing (dampened) in amplitude as a result of inevitable loss of energy.

DD. Abbreviation for a light regime of constant darkness.

Desynchonisation. (1) External: loss of synchronisation between rhythm and *Zeitgeber*. (2) Internal: loss of synchronisation between two rhythms within an organism.

Diurnal. An activity or process that occurs during the daytime (light).

Drosophila. *Drosophila melanogaster* (fruit-fly). A common model organism in genetics research, due to its short generation time and simple maintenance.

E-box. A nucleotide motif, CACGTG, involved in enhancing the expression of many clock genes.

Eclosion. Hatching of an insect pupa into an adult.

Entrainment. The process by which a biological oscillator is synchronised to an environmental rhythm such as the light/dark cycle.

Free-running. The endogenous rhythm exhibited by a circadian system under constant conditions.

Infradian rhythm. A biological rhythm with a period much longer (that is, a frequency much lower) than that of a circadian rhythm; an example is the menstrual cycle.

Jet-lag. A malaise resulting from a sudden move to a different time zone (often by trans-meridian flight).

LD. Abbreviation for a lighting regime consisting of alternating periods of light and dark.

LL. Abbreviation for a lighting regime consisting of constant light.

Masking. The phenomenon whereby an external factor directly affects the expression of an overt rhythm.

Neuropeptide. A protein hormone or messenger that is released from a nerve cell (neurosecretory cell) into the blood or intercellular spaces and changes the activity of a target cell. The surface of the target cell has specific receptors for a particular neuropeptide.

Neurospora. *Neurospora crassa* (bread mould). A filamentous fungus, used as a genetic model.

Neurotransmitter. A chemical messenger that is released from a nerve cell at its synaptic terminal and stimulates or inhibits a postsynaptic neurone by altering its electrical potential.

Opsin. The protein component of animal photopigments. Opsins use a vitamin A chromophore and possess a characteristic 'bell-shaped' absorbance spectrum.

Oscillator. A system capable of producing a regular fluctuation of an output around a mean. In chronobiology, an oscillator refers to the molecular mechanism within a cell capable of generating self-sustained rhythms.

Overt rhythm. An observable rhythm that is directly or indirectly regulated by the circadian clock.

Pacemaker. Structure capable of sustaining its own oscillations and of regulating other oscillators.

PAS domain. Protein sequence motif found in many clock proteins, involved in signalling pathways that transmit environmental information such as oxygen, redox state and light. Often associated with protein–protein interactions.

Period. The time after which a defined phase of an oscillation recurs.

Peripheral oscillator. An oscillator found in a tissue that is capable of regulating local physiology but is dependent upon a pacemaker for entrainment. Also called 'slave' oscillator.

Phase. A particular reference or reference point within the cycle of a rhythm (for example, onset of activity).

Phase shift. A single, persistent change in phase brought about by the action of a *Zeitgeber*.

Photoentrainment. The entrainment of an oscillator by the light/dark cycle.

Photoperiod. The duration of light and dark in a 24-hour, or near 24-hour, cycle.

Photopigment. A molecule that is capable of transducing the absorbance of a photon into an intracellular response.

Pineal. Neuroendocrine gland found in all vertebrates that synthesizes melatonin. Is directly photosensitive in all non-mammalian vertebrates.

SCN. Suprachiasmatic nuclei. Paired nuclei within the ventral hypothalamus that function as the circadian pacemaker in mammals.

State variable. A term used in mathematical modelling of oscillators, referring to a quantity that changes with time. More generally, a term used to denote a variable essential for defining the state of a system. Sometimes applied to the essential components (genes, proteins) required to generate a circadian oscillator.

Tau (τ). The natural period of a free-running biological rhythm.

Temperature compensation. A characteristic of circadian rhythms, whereby changes in temperature do not significantly alter the period. A Q_{10} close to 1.

Ultradian rhythm. A biological rhythm with a period much shorter (that is, a frequency much higher) than that of a circadian rhythm; an example is the heart beat.

Zeitgeber. From the German for 'time-giver', an entrainment signal.

APPENDIX I
RHYTHMS IN HUMANS

Event (High ↑; Low ↓)	00:00–02:00	02:00–04:00	04:00–06:00	06:00–08:00	08:00–10:00	10:00–12:00	12:00–14:00	14:00–16:00	16:00–18:00	18:00–20:00	20:00–22:00	22:00–24:00
PERFORMANCE												
Sleep propensity	↑									↓		
Deepest sleep			↑									
Tissue repair	↑	↑										
Labour pains start	↑											
Birth			↑									
Ovulation	↑											
Libido				↑								↑
Best chance of conception				↑								
Death			↑									
Sudden infant death			↑									
Body temperature			↓						↑	↑		
Concentration			↓			↑	↑					
Logical reasoning			↓			↑	↑					
Alertness			↓						↑	↑		
Pain intensity					↓						↑	
Skin sensitivity											↑	
Urine production		↓					↑					
Bowel movement					↑							
Heart rate					↑	↑						
Blood pressure					↑	↑						
Stomach activity											↑	↑
Heart efficiency									↑	↑		
Muscle strength									↑	↑		
Flexibility									↑	↑		
Grip strength									↑	↑		
Digestive performance					↑						↑	↑
Fat absorption		↓		↑								
Lung function								↑	↑	↑		

Event (High ↑; Low ↓)	00:00 – 02:00	02:00 – 04:00	04:00 – 06:00	06:00 – 08:00	08:00 – 10:00	10:00 – 12:00	12:00 – 14:00	14:00 – 16:00	16:00 – 18:00	18:00 – 20:00	20:00 – 22:00	22:00 – 24:00
Liver function										↑	↑	↑
Immunity											↑	↑
Allergic reactions												↑
Menopausal flushes												↑
DISEASE												
Gout	↑	↑										
Asthma		↑										
Hay fever				↑								
Rheumatoid arthritis				↑								
Migraine				↑								
Angina				↑								
Heart attack				↑	↑							
Sudden cardiac death				↑	↑							
Stroke				↑	↑							
Tooth ache								↓			↑	
Osteoarthritis									↑			
BIOCHEMISTRY												
Growth hormone	↑	↑	↑									
Progesterone	↑											
Follicle stimulating hormone						↑	↑					
Leutenising hormone	↓	↓										
Testosterone			↑	↑								
Catecholamines						↑						
Cortisol			↑									
Uric acid in the blood	↑											
Adrenaline		↓			↑				↑	↑		
Anti-inflammatory hormones		↓										
Melatonin			↑									
Serotonin			↓				↑					
Endorphins							↑				↓	
Interleukin											↑	↑

249

APPENDIX II

COPING WITH JET-LAG

You do not have to travel at jet speed to suffer from jet-lag. There is some evidence that boat-lag used to affect transatlantic passengers in the days of the large ocean liners. But the faster you cross time zones the worse it is for most people. Our biological clocks cannot cope quickly enough with the change. On average it takes about a day to adjust for each time zone crossed. The body takes about 10 days to adjust from the jet-lagging effects of crossing 10 time zones.

Part 1 Seeking out and avoiding light

There is a wide variance of response to jet-lag. Some people are badly affected, others hardly at all. Although there are no guarantees, an effective way of minimising the desynchrony caused by travelling across time zones is to use light to effectively 'shift' the clock. But light shifts the clock differently at different times of the day (Chapter 6). Light during the early night (light before the body temperature minimum at 4.00 a.m.) shifts the clock so we would start our sleep later the next night (delay). Light during the late night (light after the body temperature minimum at 4.00 a.m.) shifts the clock so that we start our sleep earlier the next night (advance) (Figure 1).

Because light at different times of the night can either advance or delay the phase of the clock, it is essential when crossing several time zones that light exposure in the new time zone should move the clock in the appropriate direction. We need to seek out, or hide from, natural sunlight reaching the retina at different times when arriving in the new time zone. A pair of dark sunglasses is the simplest means of dealing with the problems of inappropriate light exposure.

The rule about seeking out or hiding from daylight depends on the number of time zones crossed and the direction of travel.

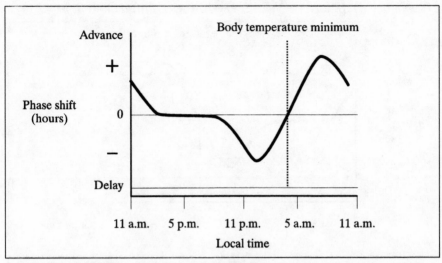

Figure 1. Diagram showing the idealised phase response curve (PRC) of a human subject. If exposed to light before the temperature minimum at about 4.00 a.m. this will have a delaying effect upon the clock. Light after the temperature minimum will advance the body clock.

Travel WEST, we need to DELAY the onset of sleep

Crossing 2–8 time zones WEST: seek out evening light

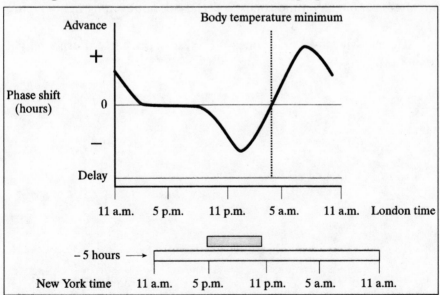

Figure 2. If travelling from London to New York (five hours behind London), then upon arrival in New York seek out the late afternoon/evening light (grey bar) for a few days. This light will fall upon the delay portion of the PRC and shift the clock in the appropriate direction.

251

Crossing 8–12 time zones WEST: seek out early afternoon light

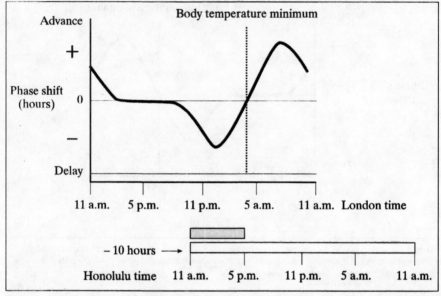

Figure 3. If travelling from London to Honolulu (10 hours behind London), then upon arrival in Honolulu seek out the late morning/afternoon light (grey bar) for several days. This light will fall upon the delay portion of the PRC and shift the clock in the appropriate direction. Avoid light in the late afternoon and early evening as this light will fall upon the advance region of the PRC.

So, after crossing 2–8 time zones WEST, seek out evening light (Figure 2).

After crossing 8–12 time zones WEST, seek out early afternoon light (Figure 3).

Travel EAST, we need to ADVANCE the onset of sleep

Crossing 2–8 time zones EAST: seek out morning light

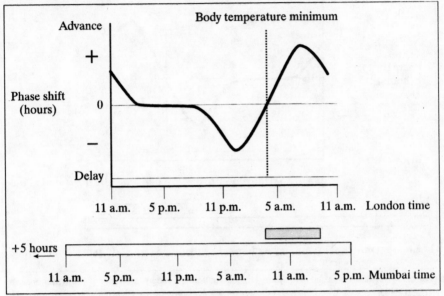

Figure 4. If travelling from London to Mumbai (five hours in front of London), then upon arrival in Mumbai seek out the morning and early afternoon light (grey bar) for several days. This light will fall upon the advance portion of the PRC and shift the clock in the appropriate direction. Avoid light of the very early morning as this light will fall upon the delay region of the PRC.

Crossing 6–12 time zones EAST: first, avoid morning light and seek out mid-afternoon or evening light for about four days; by then the clock will have advanced, so now seek out morning light

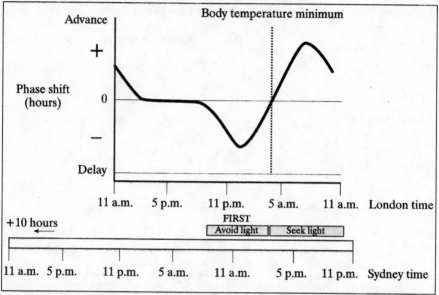

Figure 5. If travelling from London to Sydney (10 hours in front of London), then upon arrival in Sydney avoid morning light for three or four days and seek out light in the mid-afternoon. Unless morning light is avoided for the first few days after arrival, light will fall upon the delay portion of the PRC and move the clock in the wrong direction for entrainment. If you restrict your light exposure to the mid-afternoon and early evening the clock advances. But after about four to five days, light will begin to fall upon the advance portion of the PRC. As a result, you should seek out morning light after about five days of your arrival in Sydney.

So, after crossing 2–6 time zones EAST, seek out morning light (Figure 4).

After crossing 6–12 time zones EAST (Figure 5), (1) avoid morning light and seek out mid-afternoon/evening light for about four days; (2) after about four days the clock will have advanced, so now seek out morning light.

Part 2 Using melatonin

Melatonin is a natural hormone produced by the pineal during the night. It can help alleviate jet-lag, but the melatonin must be taken at a particular time of day. Taken at other times it may even be unhelpful. It is not worth using melatonin for time zone changes of less than five hours.

Melatonin in doses of about five milligrams is freely available in the USA, where it is licensed for human use. It is not available without a prescription in the UK. Compiled with advice from Prof. Josephine Arendt (University of Surrey, UK).

Travelling WEST, we need to DELAY the onset of sleep
When going west, take a capsule at local bedtime, 11.00 p.m. or later, for four days after arrival. If you wake in the very early hours of the morning (before 4.00 a.m.) you may take another capsule. Be aware that taken as late as this melatonin may make you sleepy in the morning. Do not take a capsule before your flight if going west.

Travelling EAST, we need to ADVANCE the onset of sleep
When going east, take one capsule on departure day, if necessary on the flight at the destination bedtime (for example, if flying from London to Sydney, and you fly at 10.00 a.m. from London, then take a capsule in flight at 1.30 p.m. London time, which is equivalent to 11.30 p.m. Sydney time). On arrival take a capsule at local bedtime (11.30 p.m.) for four days.

Take care: Melatonin can induce sleepiness and lowered alertness in sensitive individuals. You are advised not to drive, operate heavy or dangerous machinery, or do equivalent tasks requiring alertness, for four or five hours after taking melatonin. Long-haul pilots and crew are advised not to use melatonin, because of potential difficulties with timing the dose. If you or a close blood relative suffer from a psychiatric condition or migraine you are advised not to use melatonin. If you are under 18 years old or know (or suspect) that you are pregnant you are advised not to use melatonin. Possible side effects are sleepiness, headache (infrequent) and nausea (very infrequent).

REFERENCES

ABC (2001) *Good Morning America*, 3 December.

Aggelopoulos, N. C. & Meissl, H. (2000) Responses of neurones of the rat suprachiasmatic nucleus to retinal illumination under photopic and scotopic conditions. *J Physiol*, **523**, 211–22.

Aguilar-Roblero, R., Aréchiga, H., Ashkenazi, I., Burioka, N., Cipolla-Neto, J., Cornélissen, G., Markus, R., Marques, N., Menezes, A. A. L., Monk, T. H., Ralph, M., Valdez-Ramirez, P. & Menna-Bareto, L. (2001) The brain decade in debate: IV. Chronobiology. *Braz J Med Biol Res*, **34**, 831–841

Arendt, J., Aldhous, M. & Wright, J. (1988) Synchronisation of a disturbed sleep-wake cycle in a blind man by melatonin treatment. *Lancet*, **1**, 772–73.

Arendt, J., Deacon, S., English, J., Hampton, S. & Morgan, L. (1995) Melatonin and adjustment to phase shift. *J Sleep Res*, **4**, 74–79.

Aristophanes (1978) *The Knights, Peace, Wealth, The Birds, The Assembly Women* (Transl. D. Barrett & A. H. Sommerstein). Viking Press, New York.

Aristotle (2002) *Historia Animalium* (ed. D. M. Balme). Cambridge University Press, Cambridge.

Aschoff, J. (1960) Exogenous and endogenous components in circadian rhythms. *Cold Spring Harbor Symposia on Quantitative Biology: Biological Clocks*. Long Island Biological Association, Cold Spring Harbor, L.I., New York, The Biological Laboratory Cold Spring Harbor, pp. 11–28.

Aschoff, J. (1964) Survival value of diurnal rhythms. *Symp Zool Soc (Lond)*, **13**, 79–98.

Aschoff, J. (1965) Circadian rhythms in man. *Science*, **148**, 1427–32.

Aschoff, J., Gerecke, U. & Wever, R. (1967) Phase relations between the circadian activity periods and core temperature in humans. [In German.] *Pflügers Arch Gesamte Physiol Menschen Tiere*, **295**, 173–83.

Aschoff, J., Fatranska, M., Giedke, H., Doerr, P., Stamm, D. & Wisser, H. (1971) Human circadian rhythms in continuous darkness: entrainment by social cues. *Science*, **171**, 213–15.

Atkinson, G. & Reilly, T. (1996) Circadian variation in sports performance. *Sports Med*, **21**, 292–312.

Augustine of Hippo (1998) *Confessions* (ed. J. J. O'Donnell). Oxford University Press, Oxford.

Aveni, A. (1990) *Empires of Time*. Tauris Parke, New York.

Baldwin, S. (1986) *John Ray, Essex Naturalist*. Baldwin Books, Witham.

Balsalobre, A., Damiola, F. & Schibler, U. (1998) A serum shock induces circadian gene expression in mammalian tissue culture cells. *Cell*, **93**, 929–37.

Barlow, R. B., Jr (1990) What the brain tells the eye. *Sci Am*, **262**, 90–95.

Barnes, J. W., Tischkau, S. A., Barnes, J. A., Mitchell, J. W., Burgoon, P. W., Hickok, J. R. & Gillette, M. U. (2003) Requirement of mammalian Timeless for circadian rhythmicity. *Science* **302**, 439–442.

Batt, S. (2000) What light through window wreaks – circadian rhythms and breast cancer. *Breast Cancer Action News Letter*.

Bennett, M., Schatz, M. F., Rockwood, H. & Wiesenfeld, K. (2002) Huygens's clocks. *Proc R Soc Lond A*, **458**, 563–79.

Bernard, C. (1859) *Leçons sur les Propriétés Physiologiques et les Alterations Pathologiques de L'Organisme*. Baillière's, Paris.

Berson, D. M., Dunn, F. A. & Takao, M. (2002) Phototransduction by retinal ganglion cells that set the circadian clock. *Science*, **295**, 1070–73.

Biebach, H., Falk, H. & Krebs, J. R. (1991) The effect of constant light and phase shifts on a learned time-place association in garden warblers (*Sylvia borin*): hourglass or circadian clock? *J Biol Rhythms*, **6**, 353–65.

Bjarnason, G. A., Jordan, R. C. & Sothern, R. B. (1999) Circadian variation in the expression of cell-cycle proteins in human oral epithelium. *Am J Pathol*, **154**, 613–22.

Black, H. R., Elliott, W. J., Neaton, J. D., Grandits, G., Grambsch, P., Grimm, R. H., Jr, Hansson, L., Lacoucière, Y., Muller, J., Sleight, P., Weber, M. A., White, W. B., Williams, G., Wittes, J., Zanchetti, A., Fakouhi, T. D. & Anders, R. J. (2001) Baseline characteristics and early blood pressure control in the CONVINCE trial. *Hypertension*, **37**, 12–18.

Black, H. R., Elliott, W. J., Grandits, G., Grambsch, P., Lucente, T., White, W. B., Neaton, J. D., Grimm, R. H. Jr, Hansson, L., Lacourcière, Y., Muller, J., Sleight, P., Weber, M. A., Williams, G., Wittes, J., Zanchetti, A. & Anders, R. J. (2003) Principal results of the Controlled Onset Verapamil Investigation of Cardiovascular End Points (CONVINCE) trial. *JAMA*, **289**, 2073–82.

Bolles, R. C. & Stokes, L.W. (1965) Rat's anticipation of diurnal and a-diurnal feeding. *J Comp Physiol Psychol*, **60**, 290–94.

Borbely, A. A. (1982) A two process model of sleep regulation. *Hum Neurobiol*, **1**, 195–204.

Bos, J.L. (1989) *ras* oncogenes in human cancer: a review. *Cancer Res*, **49**, 4682–89.

Brady, J. (1979) *Biological Clocks*. Edward Arnold, London.

Brady, J. (1982) *Biological Timekeeping*. Cambridge University Press, Cambridge.

Brandstatter, R. (2002) The circadian pacemaking system of birds. In Kumar, V. (ed.) *Biological Rhythms*. Narose Publishing House, New Delhi.

Broughton, R. J. (1998) SCN controlled circadian arousal and the afternoon 'nap zone'. *Sleep Res Online*, **1**, 166–78.

Brown, J. F. A. (1960) Response to pervasive geophysical factors and the biological clock problem. *Cold Spring Harbor Symposia on Quantitative Biology: Biological Clocks*. Long Island Biological Association, Cold Spring Harbor, L.I., New York, The Biological Laboratory Cold Spring Harbor, pp. 57–71.

Buguet, A., Moroz, D. E. & Radomski, M. W. (2003) Modafinil – medical considerations for use in sustained operations. *Aviat Space Environ Med*, **74**, 659–63.

Buijs, R. M. & Kalsbeek, A. (2001) Hypothalamic integration of central and peripheral clocks. *Nat Rev Neurosci*, **2**, 521–26.

Bunk, S. (2002) Big genes are back. *Scientist*, **16**, 24–29.

Bünning, E. (1973) *The Physiological Clock – Circadian Rhythms in Biological Chronometry*. The English Universities Press Ltd, London.

Burns, E. R. (2000) Biological time and *in vivo* research: a field guide to pitfalls. *Anat Rec*, **261**, 141–52.

Buysse, D. J., Barzansky, B., Dinges, D., Hogan, E., Hunt, C. E., Owens, J., Rosekind, M., Rosen, R., Simon, F., Veasey, S. & Wiest, F. (2003) Sleep, fatigue, and medical training: setting an agenda for optimal learning and patient care. *Sleep*, **26**, 218–25.

Cahill, G. M. (2002) Clock mechanisms in zebrafish. *Cell Tissue Res*, **309**, 27–34.

Cajochen, C., Zeitzer, J. M., Czeisler, C. A. & Dijk, D. J. (2000) Dose-response relationship for light intensity and ocular and electroencephalographic correlates of human alertness. *Behav Brain Res*, **115**, 75–83.

Cambrosio, A. & Keating, P. (1983) The disciplinary stake: the case of chronobiology. *Soc Stud Sci*, **13**, 323–53.

Camhi, J. M. (1984) *Neuroethology: Nerve Cells and the Natural Behavior of Animals*. Sinauer Associates, Sunderland, Massachusetts.

Cannon, W. B. (1947) *The Wisdom of the Body*. Kegan Paul, London.

Carrier, J. & Monk, T. H. (2000) Circadian rhythms of performance: new trends. *Chronobiol Int*, **17**, 719–32.

Chew, M. K. & Laubichler, M. D. (2003) Perceptions of science. Natural enemies – metaphor or misconception? *Science*, **301**, 52–53.

Cho, K., Ennaceur, A., Cole, J. C. & Suh, C. K. (2000) Chronic jet lag produces cognitive deficits. *J Neurosci*, **20**, RC66.

Cockell, C. S. & Rothschild, L. J. (1999) The effects of UV radiation A and B on diurnal variation in photosynthesis in three taxonomically and ecologically diverse microbial mats. *Photochem Photobiol*, **69**, 203–10.

Cohen, R. A. & Albers, H. E. (1991) Disruption of human circadian and cognitive regulation following a discrete hypothalamic lesion: a case study. *Neurology*, **41**, 726–29.

Cole, R. J., Kripke, D. F., Wisbey, J., Mason, W. J., Gruen, W., Hauri, P. J. & Juarez, S. (1995) Seasonal variation in human illumination exposure at two different latitudes. *J Biol Rhythms*, **10**, 324–34.

Cromie, W. J. (1999) Human biological clock set back an hour. *Harvard University Gazette*.

Czeisler, C. A., Shanahan, T. L., Klerman, E. B., Martens, H., Brotman, D. J., Emens, J. S., Klein, T. & Rizzo, J. F., III (1995) Suppression of melatonin secretion in some blind patients by exposure to bright light. *N Engl J Med*, **332**, 6–11.

Czeisler, C. A., Duffy, J. F., Shanahan, T. L., Brown, E. N., Mitchell, J. F., Rimmer, D. W., Ronda, J. M., Silva, E. J., Allan, J. S., Emens, J. S., Dijk, D. J. & Kronauer, R. E. (1999) Stability, precision, and near-24-hour period of the human circadian pacemaker. *Science*, **284**, 2177–81.

Darwin, C. & Darwin, F. (1880) *The Power of Movement in Plants*. John Murray, London.

David-Gray, Z. K., Janssen, J. W., DeGrip, W. J., Nevo, E. & Foster, R. G. (1998) Light detection in a 'blind' mammal. *Nat Neurosci*, **1**, 655–56.

Dawkins, R. (1976) *The Selfish Gene*. Oxford University Press, Oxford.

Dawson, A., Goldsmith, A. R., Nicholls, T. J. & Follett, B. K. (1986) Endocrine changes associated with the termination of photorefractoriness by short daylengths and thyroidectomy in starlings (*Sturnus vulgaris*). *J Endocrinol*, **110**, 73–79.

de Candolle, M. (1832) De l'influence de la lumière sur les végétaux. *Physiol Vég*, **4**, 1069.

de Mairan, J. J. O. (1729) Observation botanique. *Histoire de l'Académie Royale des Sciences*, 35–36.

DeCoursey, P. J., Krulas, J. R., Mele, G. & Holley, D. C. (1997) Circadian performance of suprachiasmatic nuclei (SCN)-lesioned antelope ground squirrels in a desert enclosure. *Physiol Behav*, **62**, 1099–108.

DeCoursey, P. J., Walker, J. K. & Smith, S. A. (2000) A circadian pacemaker in free-living chipmunks: essential for survival? *J Comp Physiol A*, **186**, 169–80.

Deinlein, M. (1997) Have wings, will travel: avian adaptations to migration. Migratory Bird Center (Smithsonian National Zoological Park), Washington DC. Fact sheet 4.

Dement, W. C. & Vaughan, C. (1999) *The Promise of Sleep*. Macmillan, London.

Dijk, D.-J. (1996) Internal rhythms in humans. *Semin Cell Dev Biol*, **7**, 831–36.

Dijk, D. J. & Czeisler, C. A. (1994) Paradoxical timing of the circadian rhythm of sleep propensity serves to consolidate sleep and wakefulness in humans. *Neurosci Lett*, **166**, 63–68.

Dijk, D. J., Neri, D. F., Wyatt, J. K., Ronda, J. M., Riel, E., Ritz-De Cecco, A., Hughes, R. J., Elliott, A. R., Prisk, G. K., West, J. B. & Czeisler, C. A. (2001) Sleep, performance, circadian rhythms, and light-dark cycles during two space shuttle flights. *Am J Physiol Regul Integr Comp Physiol*, **281**, R1647–64.

du Nouy, L. C. (1937) *Biological Time*. Macmillan, New York.

Dunbar, R. (1995) *The Trouble with Science*. Faber & Faber, London.

Dunlap, J. C. (1999) Molecular bases for circadian clocks. *Cell*, **96**, 271–90.

Edery, I. (2000) Circadian rhythms in a nutshell. *Physiol Genomics*, **3**, 59–74.

Edgar, D. M., Dement, W. C. & Fuller, C. A. (1993) Effect of SCN lesions on sleep in squirrel monkeys: evidence for opponent processes in sleep-wake regulation. *J Neurosci*, **13**, 1065–79.

Elliott, W. J. (2001) Timing treatment to the rhythm of disease. A short course in chronotherapeutics. *Postgrad Med*, **110**, 119–22, 125–26, 129.

Fischer-Perroudon, C., Mouret, J. & Jouvet, M. (1974) Case of agrypnia (4 months without sleep) in Morvan's disease. Favorable action of 5-hydroxytryptophan. *Electroencephalogr Clin Neurophysiol*, **36**, 1–18.

Folkard, S. (1983) Diurnal variation. In Hockey, R. (ed.) *Stress and Fatigue in Human Performance*. John Wiley and Sons, New York, pp. 245–72.

Folkard, S. (2000) *Shift-work and Health*. European Foundation for the Improvement of Living and Working Conditions, Dublin.

Follett, B. K. (1985) The environment and reproduction. In Austin, C. R. & Short, R. V. (eds) *Reproduction in Mammals: 4. Reproductive Fitness*. Cambridge University Press, Cambridge, pp. 103–32.

Follett, B. K. & Sharp, P. J. (1969) Circadian rhythmicity in photoperiodically induced gonadotrophin release and gonadal growth in the quail. *Nature*, **223**, 968–71.

Forel, A. (1910) *Das Sinnesleben der Insekten*. Reinhardt, München.

Foster, R. G. (1998) Photoentrainment in the vertebrates: a comparative analysis. In Lumsden, P. J. & Millar, A. J. (eds) *Biological Rhythms and Photoperiodism in Plants*. BIOS, Oxford, pp. 135–49.

Foster, R. G. & Hankins, M. W. (2002) Non-rod, non-cone photoreception in the vertebrates. *Prog Retinal Eye Res*, **21**, 507–27.

Foster, R. G. & Soni, B. G. (1998) Extraretinal photoreceptors and their regulation of temporal physiology. *Rev of Reprod*, **3**, 145–50.

Fox Keller, E. (2001) *Century of the Gene*. Harvard University Press, Boston.

Fraser, J. T. (1987) *Time: The Familiar Stranger*. University of Massachusetts Press, Amherst.

Froy, O., Gotter, A. L., Casselman, A. L. & Reppert, S. M. (2003) Illuminating the circadian clock in monarch butterfly migration. *Science*, **300**, 1303–05.

Fu, L., Pelicano, H., Liu, J., Huang, P. & Lee, C. (2002) The circadian gene Period2 plays an important role in tumor suppression and DNA damage response in vivo. *Cell*, **111**, 41–50.

Gale, C. & Martyn, C. (1998) Larks and owls and health, wealth, and wisdom. *Br Med J*, **317**, 1675–77.

Garcia, C. R., Markus, R. P. & Madeira, L. (2001) Tertian and quartan fevers: temporal regulation in malarial infection. *J Biol Rhythms*, **16**, 436–43.

Giebultowicz, J. H. & Hege, D. M. (1997) Circadian clock in malpighian tubules. *Nature*, **386**, 664–65.

Glasser, S. P., Neutel, J. M., Gana, T. J. & Albert, K. S. (2003) Efficacy and safety of a once daily graded-release diltiazem formulation in essential hypertension. *Am J Hypertens*, **16**, 51–58.

Goddard, P. (1998) *Paul Dirac: The Man and His Work*. Cambridge University Press, Cambridge.

Goldman, B. D., Darrow, J. M. & Yogev, L. (1984) Effects of timed melatonin infusions on reproductive development in the Djungarian hamster (*Phodopus sungorus*). *Endocrinology*, **114**, 2074–83.

Golombek, D. A. & Yannielli, P. C. (1996) Chronoliterature: biological rhythms in Argentine fiction. *Chronobiol Int*, **13**, 487–88.

Gould, J. L. & Gould, C. G. (1999) *The Animal Mind*. W. H. Freeman, New York.

Greene, B. (1999) *The Elegant Universe*. Vintage, London.

Griffin, D. R. (1964) *Bird Migration*. The Natural History Press, New York.

Groopman, J. (2001) Eyes wide open. *New Yorker*, 3 December, 52–57.

Gwinner, E. (1996a) Circannual clocks in avian reproduction and migration. *Ibis*, **138**, 47–63.

Gwinner, E. (1996b) Circadian and circannual programmes in avian migration. *J Exp Biol*, **199**, 39–48.

Halberg, F. (1976) Chronobiology in 1975. *Chronobiologia*, **3**, 1–11.

Halberg, F., Cornelissen, G., Otsuka, K., Katinas, G. & Schwartzkopff, O. (2001) Essays on chronomics spawned by transdisciplinary chronobiology. Witness in time: Earl Elmer Bakken. *Neuroendocrinol Lett*, **22**, 359–84.

Hall, J. C. (2003) Genetics and molecular biology of rhythms in *Drosophila* and other insects. *Adv Genet*, **48**, 1–280.

Hardin, P. E., Hall, J. C. & Rosbash, M. (1990) Feedback of the *Drosophila* period gene product on circadian cycling of its messenger RNA levels. *Nature*, **343**, 536–40.

Harkness, J. A., Richter, M. B., Panayi, G. S., Van de Pette, K., Unger, A., Pownall, R. & Geddawi, M. (1982) Circadian variation in disease activity in rheumatoid arthritis. *Br Med J (Clin Res Ed)*, **284**, 551–54.

Harper, D.G., Stopa, E. G., McKee, A. C., Satlin, A., Harlan, P. C., Goldstein, R. & Volicer, L. (2001) Differential circadian rhythm disturbances in men with Alzheimer disease and frontotemporal degeneration. *Arch Gen Psychiatry*, **58**, 353–60.

Hastings, J. W. (2001) Fifty years of fun. *J Biol Rhythms*, **16**, 5–18.

Hastings, J. W. & Sweeney, B. M. (1958) A persistent diurnal rhythm on luminescence in *Gonyaulax polyedra*. *Biol Bull*, **115**, 440–458.

Hattar, S., Lucas, R. J., Mrosovsky, N., Thompson, S., Douglas, R. H., Hankins, M. W., Lem, J., Biel, M., Hofmann, F., Foster, R. G. & Yau, K. W. (2003) Melanopsin and rod-cone photoreceptive systems account for all major accessory visual functions in mice. *Nature*, 424, 76–81.

Hauser, M. (2000) *Wild Minds*. Penguin, London.

Hendricks, J. C. (2003) Genetic models in applied physiology: invited review. Sleeping flies don't lie: the use of *Drosophila melanogaster* to study sleep and circadian rhythms. *J Appl Physiol*, **94**, 1660–72.

Herodotus (2003) *The Histories, Book II* (ed. J. M. Marincola & A. de Selincourt). Penguin, London.

Highkin, H. R. & Hanson, J. B. (1954) Possible interactions between light-dark cycles and endogenous daily rhythms on the growth of tomato plants. *Plant Physiol*, **29**, 301–02.

Hodge, W. (2001) Reindeer herders at home on a very cold range. *New York Times International*, 26 March.

Homer (1991) *The Iliad* (ed. B. Knox). Penguin, London.

Hoyert, D. L., Kochanek, K. D. & Murphy, S. L. (1999) Deaths: final data for 1997. *Natl Vital Stat Rep*, **47**, 1–104.

Hrushesky, W. J. (1985) Circadian timing of cancer chemotherapy. *Science*, **228**, 73–75.

Hrushesky, W. J. (1991) The multifrequency (circadian, fertility cycle, and season) balance between host and cancer. *Ann N Y Acad Sci*, **618**, 228–56.

Hutchenson, C. (1999) *New York Times*, 14 December.

Huygens, C. (1893) *Oeuvres complètes de Christiaan Huygens*. Martinus Nijhoff, The Hague.

Inouye, S. T. & Kawamura, H. (1979) Persistence of circadian rhythmicity in a mammalian hypothalamic 'island' containing the suprachiasmatic nucleus. *Proc Nat Acad Sci USA*, **76**, 5962–66.

Jacob, F. (1994) *The Possible and the Actual*. University of Washington Press, Seattle.

Johnson, C., Knight, M., Trewavas, A. & Kondo, T. (1998) A clockwork green: circadian programmes in photosynthetic organisms. In Lumsden, P. J. & Millar, A. J. (eds) *Biological Rhythms and Photoperiodism in Plants*. BIOS, Oxford, pp. 1–34.

Johnson, C. H. & Golden, S. S. (1999) Circadian programs in cyanobacteria: adaptiveness and mechanism. *Annu Rev Microbiol*, **53**, 389–409.

Karban, R., Black, C. A. & Weinbaum, S. A. (2000) How 17-year cicadas keep track of time. *Ecology Lett*, **3**, 253–56.

Klein, D. C. (1993) The mammalian melatonin rhythm generating system. In Wetterberg, L. (ed.) *Light and Biological Rhythms in Man*. Pergamon Press, Oxford, pp. 55–71.

Kleitman, N. (1963) *Sleep and Wakefulness*. University of Chicago Press, Chicago.

Kloner, R. A., Poole, W. K. & Perritt, R. L. (1999) When throughout the year is coronary death most likely to occur? A 12-year population-based analysis of more than 220,000 cases. *Circulation*, **100**, 1630–34.

Kmietowicz, Z. (1997) Chemotherapy better tolerated when matched to body's rhythm. *Br Med J*, **315**, 623–28.

Koltermann, R. (1974) Periodicity in the activity and learning performance of the honey bee. In Browne, L. B. (ed.) *Experimental Analysis of Insect Behaviour*. Springer-Verlag, Berlin, pp. 218–27.

Kondo, T. (1998) Circadian oscillator of prokaryote: circadian timing by Kai clock proteins and associates in cyanobacteria. In Honma, K., Honma, S. (eds) *Zeitgebers, Entrainment and Masking of the Circadian System*. Hokkaido University Press, Sapporo, pp. 267–79.

Kondo, T., Strayer, C. A., Kulkarni, R. D., Taylor, W., Ishiura, M., Golden, S. S. & Johnson, C. H. (1993) Circadian rhythms in prokaryotes: luciferase as a reporter of circadian gene expression in cyanobacteria. *Proc Natl Acad Sci USA*, **90**, 5672–76.

Konopka, R. J. & Benzer, S. (1971) Clock mutants of *Drosophila melanogaster*. *Proc Natl Acad Sci USA*, **68**, 2112–16.

Kramer, G. (1949) Über Richtungstendenzen bei der nächtlichen Zugunruhe gekäfigter Vögel. In Mayr, E. & Schüz, E. (eds) *Ornithologie als Biologische Wissenschaft*. Winter, Heidelberg, pp. 269–83.

Kramer, G. (1952) Experiments on bird orientation. *Ibis*, **94**, 265–85.

Kreitzman, L. (1999) *The 24 Hour Society*. Profile Books, London.

Kripke, D. F., Garfinkel, L., Wingard, D. L., Klauber, M. R. & Marler, M. R. (2002) Mortality associated with sleep duration and insomnia. *Arch Gen Psychiatry*, **59**, 131–36.

Kruszelnicki, K. S. (1998) Rhythm of life. *Great Moments in Science*. ABC, Sydney. <http:///www.abc.net.au/science/k2/moments/gmis9834.htm>

Lakin-Thomas, P. L. (2000) Circadian rhythms: new functions for old clock genes. *Trends Genet*, **16**, 135–42.

Landes, D. (2000) *Revolution in Time*. Viking, New York.

Lévi, F., Giacchetti, S., Zidani, R., Brezault-Bonnet, C., Tigaud, J. M., Goldwasser, F. & Misset, J. L. (2001) Chronotherapy of colorectal cancer metastases. *Hepatogastroenterology*, **48**, 320–22.

Lewis-Williams, D. (2002) *The Mind in the Cave: Consciousness and the Origins of Art*. Thames & Hudson, London.

Lincoln, G. (1999) Melatonin modulation of prolactin and gonadotrophin secretion. Systems ancient and modern. *Adv Exp Med Biol*, **460**, 137–53.

Lockley, S. W., Skene, D. J., Arendt, J., Tabandeh, H., Bird, A. C. & Defrance, R. (1997) Relationship between melatonin rhythms and visual loss in the blind. *J Clin Endocrinol Metab*, **82**, 3763–70.

Loehle, C. (1990) A guide to increased creativity in research? Inspiration or perspiration? *BioScience*, **40**, 123–29.

Loewe, M. (1999) Cyclical and linear concepts of time in China. In Lippincott, K. (ed.) *The Story of Time*. Merrell Holberton, London, pp. 76–79.

Lofts, B. (1970) *Animal Photoperiodism*. Edward Arnold, London.

Magnusson, A. & Boivin, D. (2003) Seasonal affective disorder: an overview. *Chronobiol Int*, **20**, 189–207.

Manning, A. & Stamp-Dawkins, M. (1998) *Introduction to Animal Behaviour*. Cambridge University Press, Cambridge.

Margulis, L. (1998) *The Symbiotic Planet: A New Look at Evolution*. Basic Books, New York.

Marvell, M. (1681) *Miscellaneous Poems* (facsimile edition, 1969). Scolar Press, Marston.

McDonald, E. R., 3rd & El-Deiry, W. S. (2000) Cell cycle control as a basis for cancer drug development (Review). *Int J Oncol*, **16**, 871–86.

Menaker, M. (1972) Nonvisual light reception. *Sci Am*, March, 22–29.

Menaker, M. & Binkley, S. (1981) Neural and endocrine control of circadian rhythms in the vertebrates. In Aschoff, J. (ed.) *Handbook of Behavioral Neurobiology*. Plenum Press, New York, pp. 243–55.

Menaker, M. & Tosini, G. (1996) The evolution of vertebrate circadian systems. In Honma, K. & Honma, S. (eds) *Circadian Organization and Oscillatory Coupling: Proceedings of the sixth Sapporo Symposium on*

Biological Rhythms. Hokkaido University Press, Sapporo, Japan, pp. 39–52.

Merrow, M., Brunner, M. & Roenneberg, T. (1999) Assignment of circadian function for the *Neurospora* clock gene frequency. *Nature*, **399**, 584–86.

Mignot, E., Taheri, S. & Nishino, S. (2002) Sleeping with the hypothalamus: emerging therapeutic targets for sleep disorders. *Nat Neurosci*, **5 Suppl**, 1071–75.

Millar, A. J. (1998) The cellular organisation of circadian rhythms in plants: not one but many clocks. In Lumsden, P. J. & Millar, A. J. (eds) *Biological Rhythms and Photoperiodism in Plants*. Oxford BIOS Scientific, Oxford, pp. 51–68.

Mohawk, J. (2000) <http://news.bbc.co.uk/hi/english/static/special_report/1999/12/99/back_to_the_future/john_mohawk.stm>

Moore, J. (2002) Controlled onset, extended release verapamil similar to standard of care. *Today in Cardiology*, July.

Moore, R. Y. & Eichler, V. B. (1972) Loss of a circadian adrenal corticosterone rhythm following suprachiasmatic lesions in the rat. *Brain Res.*, **42**, 201–06.

Moore-Ede, M. C. (1993) *The 24 hour Society*. Addison-Wesley, New York.

Moore-Ede, M. C., Sulzman, F. M. & Fuller, C. A. (1982) *The Clocks that Time Us: Physiology of the Circadian Timing System*. Harvard University Press, Cambridge, Massachusetts.

Moran, T. H. & Schulkin, J. (2000) Curt Richter and regulatory physiology. *Am J Physiol Regul Integr Comp Physiol*, **279**, R357–63.

Morell, V. (1996) Setting a biological stopwatch. *Science*, **271**, 905–06.

Mori, T. & Johnson, C. H. (2001) Circadian programming in cyanobacteria. *Semin Cell Dev Biol*, **12**, 271–78.

Morris, S. (2003) How have the coalition troops kept going on so little sleep? *The Guardian*, London, 10 April.

Moss, L. (2003) *What Genes Can't Do*. MIT Press, Boston.

Mrosovsky, N. (1990) *Rheostasis: The Physiology of Change*. Oxford University Press, New York and Oxford.

National Commission on Sleep Disorders Research (1992) *Wake Up America: A National Sleep Alert*. Department of Health and Human Services, Bethesda, Maryland.

Nowotny, H. (1994) *Time – The Modern and Postmodern Experience*. Polity Press, Cambridge.

Okawa, M., Uchiyama, M., Ozaki, S., Shibui, K. & Ichikawa, H. (1998) Circadian rhythm sleep disorders in adolescents: clinical trials of combined treatments based on chronobiology. *Psychiatry Clin Neurosci*, **52**, 483–490.

Oklejewicz, M., Hut, R. A., Daan, S., Loudon, A. S. I. & Stirland, J. A. (1997) Metabolic rate changes proportionally to circadian frequency in tau mutant hamsters. *J Biol Rhythms*, **12**, 413–22.

Olden, K. & Wilson, S. (2000) Environmental health and genomics: visions and implications. *Nat Rev Genet*, **1**, 149–53.

Pace-Schott, E. F. & Hobson, J. A. (2002) The neurobiology of sleep: genetics, cellular physiology and subcortical networks. *Nat Rev Neurosci*, **3**, 591–605.

Palevitz, B. A. (2002) Flower of a find. *Scientist*, **16**, 23–24.

Pengelley, E. T. & Fisher, K. C. (1966) Locomotor activity patterns and their relation to hibernation in the golden-mantled ground squirrel. *J Mammal*, **47**, 63–73.

Pittendrigh, C. S. (1960) Circadian rhythms and the circadian organisation of living systems. *Cold Spring Harbor Symposia on Quantitative Biology: Biological Clocks*. Long Island Biological Association, Cold Spring Harbor, L.I., New York, The Biological Laboratory Cold Spring Harbor, pp. 159–84.

Pittendrigh, C. S. (1993) Temporal organization: reflections of a Darwinian clock-watcher. *Annu Rev Physiol*, **55**, 17–54.

Preston, R. (2000) The genome warrior. *The New Yorker*, 12 June.

Prigogine, I. & Stengers, I. (1984) *Order Out of Chaos*. Flamingo, London.

Provencio, I., Rodriguez, I. R., Jiang, G., Hayes, W. P., Moreira, E. F. & Rollag, M. D. (2000) A novel human opsin in the inner retina. *J Neurosci*, **20**, 600–605.

Ptácek, L. J. (2001) First human circadian rhythm gene identified. *Howard Hughes Medical Institute Research News*, 12 January.

Rajaratnam, S. M. & Arendt, J. (2001) Health in a 24-h society. *Lancet*, **358**, 999–1005.

Ralph, M. R. & Menaker, M. (1988) A mutation of the circadian system in golden hamsters. *Science*, **241**, 1225–27.

Ralph, M. R., Foster, R. G., Davis, F. C. & Menaker, M. (1990) Transplanted suprachiasmatic nucleus determines circadian period. *Science*, **247**, 975–78.

Recht, L. D., Lew, R. A. & Schwartz, W. J. (1995) Baseball teams beaten by jet lag. *Nature*, **377**, 583.

Redman, J., Armstrong, S. & Ng, K. T. (1983) Free-running activity rhythms in the rat: entrainment by melatonin. *Science*, **219**, 1089–91.

Reierth, E. & Stokkan, K.-A. (2002) Biological rhythms in Arctic animals. In Kumar, V. (ed.) *Biological Rhythms*. Narosa Publishing House, New Delhi, pp. 216–23.

Reierth, E., Van't Hof, T. J. & Stokkan, K. A. (1999) Seasonal and daily variations in plasma melatonin in the high-Arctic Svalbard ptarmigan (*Lagopus mutus hyperboreus*). *J Biol Rhythms*, **14**, 314–19.

Reiter, R. J. (1975) Exogenous and endogenous control of the annual reproductive cycle in the male golden hamster: participation of the pineal gland. *J Exp Zool*, **191**, 111–20.

Rensing, L., Meyer-Grahle, U. & Ruoff, P. (2001) Biological timing and the clock metaphor: oscillatory and hourglass mechanisms. *Chronobiol Int*, **18**, 329–69.

Reppert, S. M. (1997) Melatonin receptors: molecular biology of a new family of G protein-coupled receptors. *J Biol Rhythms*, **12**, 528–31.

Richter, C. (1967) Sleep and activity: their relation to the 24 hour clock. In Kety, S. S., Evarts, E. V. & Williams, H. L. (eds) *Sleep and Altered States of Consciousness*. Williams & Wilkins Company, Baltimore, pp. 8–29.

Richter, C. (1985) It's a long way to Tipperary: land of my genes. In Dewsdury, D.A. (ed.) *Leaders in the Study of Animal Behavior*. Bucknell University Press, Lewisburg, Pennsylvania.

Rivard, G. E., Infante-Rivard, C., Dresse, M. F., Leclerc, J. M. & Champagne, J. (1993) Circadian time-dependent response of childhood lymphoblastic leukemia to chemotherapy: a long-term follow-up study of survival. *Chronobiol Int*, **10**, 201–04.

Roenneberg, T. & Merrow, M. (2001) The role of feedback in circadian systems. In Honma, K. & Honma, S. (eds) *Zeitgebers, Entrainment and Masking of the Circadian System*. Hokkaido University Press, Sapporo, pp. 113–29.

Rose, S. (1998) *Lifelines*. Penguin, London.

Rossi, B., Zani, A. & Mecacci, L. (1983) *Percept Mot Skills*, **57**, 27–30.

Rossi, E. L. (1992) *Ultradian Rhythms in Life Processes: An Inquiry into Fundamental Principles of Chronobiology and Psychobiology*. Springer-Verlag Palisades Gateway Publishing, New York.

Rothschild, L. (1998) Protists, UV and evolution. <http://www.accessexcellence.org/BF/bf05/rothschild/bf05b1.html>

Rothschild, L. J. & Cockell, C. S. (1999) Radiation: microbial evolution, ecology, and relevance to mars missions. *Mutat Res*, **430**, 281–91.

Rowan, W. (1925) Relation of light to bird migration and developmental changes. *Nature*, **115**, 494–95.

Rowan, W. (1929) Experiments in bird migration I: manipulation of the reproductive cycle. *Proc Boston Soc Nat Hist*, **39**, 151–208.

Rozin, P. (1976). In Blass, E. (ed.) *The Psychobiology of Curt Richter*. York Press, Baltimore, pp. xv–xxix.

Russo, E. (1999) Circadian Rhythms. *Scientist*, **13**, 16.

Sacks, O. (1991) *Awakenings*. Picador, London.

Saper, C. B., Chou, T. C. & Scammell, T. E. (2001) The sleep switch: hypothalamic control of sleep and wakefulness. *Trends Neurosci*, **24**, 726–31.

Scaravilli, F., Cordery, R. J., Kretzschmar, H., Gambetti, P., Brink, B., Fritz, V., Temlett, J., Kaplan, C., Fish, D., An, S. F., Schulz-Schaeffer, W. J. & Rossor, M. N. (2000) Sporadic fatal insomnia: a case study. *Ann Neurol*, **48**, 665–68.

Schäfer, E. A. (1907) On the incidence of daylight as a determining factor in bird migration. *Nature*, **77**, 159–63.

Scheuerlein, A. & Gwinner, E. (2002) Is food availability a circannual zeitgeber in tropical birds? A field experiment on stonechats in tropical Africa. *J Biol Rhythms*, **17**, 171–80.

Schmidt-Nielsen, K., Schmidt-Nielsen, B., Jarnum, S. A. & Houpt, T. R. (1957) Body temperature of the camel and its relation to water economy. *Am J Physiol*, **188**, 103–12.

Schopf, W. (1999) *The Cradle of Life*. Princeton University Press, Princeton.

Schrödinger, E. (1992) *What Is Life?: With Mind and Matter and Autobiographical Sketches (CANTO)*. Cambridge University Press, Cambridge.

Schwartz, W. J., Davidsen, L. C. & Smith, C. B. (1980) In vivo metabolic activity of a putative circadian oscillator, the rat suprachiasmatic nucleus. *J Comp Neurol*, **189**, 157–67.

Sehgal, A., Rothenfluh-Hilfiker, A., Hunter-Ensor, M., Chen, Y., Myers, M. P. & Young, M. W. (1995) Rhythmic expression of timeless: a basis for promoting circadian cycles in period gene autoregulation. *Science*, **270**, 808–10.

Sekaran, S., Foster, R. G., Lucas, R. J. & Hankins, M. W. (2003) Calcium imaging reveals a network of intrinsically light-sensitive inner-retinal neurons. *Curr Biol*, **13**, 1290–98.

Shettleworth, S. J. (1998) *Cognition, Evolution and Behaviour*. Oxford University Press, Oxford.

Silver, R., Lesauter, J., Tresco, P. A. & Lehman, M. N. (1996) A diffusible coupling signal from the transplanted suprachiasmatic nucleus controlling circadian locomotor rhythms. *Nature*, **382**, 810–13.

Simpson, S. & Galbraith, J. J. (1905) An investigation into the diurnal variation of the body temperature of nocturnal and other birds and a few mammals. *J Physiol (Lond)*, **33**, 225–38.

Singer, C. M. & Lewy, A. J. 1999 Does our DNA determine when we sleep? *Nat Med*, **5**, 983.

Siwicki, K. K., Eastman, C., Petersen, G., Rosbash, M. & Hall, J. C. (1988) Antibodies to the period gene product of *Drosophila* reveal diverse tissue distribution and rhythmic changes in the visual system. *Neuron*, **1**, 141–50.

Smolensky, M. H. (1998) Knowledge and attitudes of American physicians and public about medical chronobiology and chronotherapeutics. Findings of two 1996 Gallup surveys. *Chronobiol Int*, **15**, 377–94.

Smolensky, M. H. & Lamberg, L. (2000) *The Body Clock Guide to Better Health: How to Use Your Body's Natural Clock to Fight Illness and Achieve Maximum Health*. Henry Holt and Company, New York.

Somers, D. E., Devlin, P. F. & Kay, S. A. (1998) Phytochromes and cryptochromes in the entrainment of the *Arabidopsis* circadian clock. *Science*, **282**, 1488–90.

Soni, B. G., Philp, A. R., Knox, B. E. & Foster, R. G. (1998) Novel retinal photoreceptors. *Nature*, **394**, 27–28.

Spaceflight Now (2002) Humans' internal clock not ready for Mars time (National Space Biomedical Research Institute News Release, 24 January). <http://spaceflightnow.com/news/n0201/24marstime/>

Stephan, F. K. & Zucker, I. (1972) Circadian rhythms in drinking behavior and locomotor activity of rats are eliminated by hypothalamic lesions. *Proc Natl Acad Sci USA*, **69**, 1583–86.

Stevens, R. G. & Rea, M. S. (2001) Light in the built environment: potential role of circadian disruption in endocrine disruption and breast cancer. *Cancer Causes Control*, **12**, 279–87.

Stokkan, K. A., Yamazaki, S., Tei, H., Sakaki, Y. & Menaker, M. (2001) Entrainment of the circadian clock in the liver by feeding. *Science*, **291**, 490–93.

Storch, K. F., Lipan, O., Leykin, I., Viswanathan, N., Davis, F. C., Wong, W. H. & Weitz, C. J. (2002) Extensive and divergent circadian gene expression in liver and heart. *Nature*, **417**, 78–83.

Toh, K. L., Jones, C. R., He, Y., Eide, E. J., Hinz, W. A., Virshup, D. M., Ptácek, L. J. & Fu, Y. H. (2001) An hPer2 phosphorylation site mutation in familial advanced sleep phase syndrome. *Science*, **291**, 1040–43.

UNEP (1995) *Global Biodiversity Assessment*. Cambridge University Press, Cambridge.

Van Someren, E. J., Swaab, D. F., Colenda, C. C., Cohen, W., McCall, W. V. & Rosenquist, P. B. (1999) Bright light therapy: improved sensitivity to its effects on rest-activity rhythms in Alzheimer patients by application of nonparametric methods. *Chronobiol Int*, **16**, 505–18.

Vitaterna, M. H., King, D. P., Chang, A., Kornhauser, J. M., Lowrey, P. L., McDonald, J. D., Dove, W. F., Pinto, L. H., Turek, F. W. & Takahashi, J. S. (1994) Mutagenesis and mapping of a mouse gene, clock, essential for circadian behavior. *Science*, **264**, 719–25.

von Economo, C. (1931) Sleep as a problem of localisation. *J Nerv Mental Dis*, **71**, 249–69.

von Frisch, K. (1953) *The Dancing Bees*. Harcourt, Brace, Jovanovich, New York.

Wayne, N. L., Malpaux, B. & Karsch, F. J. (1988) How does melatonin code for day length in the ewe: duration of nocturnal melatonin release or coincidence of melatonin with a light-entrained sensitive period? *Biol Reprod*, **39**, 66–75.

Weinberg, S. (1998) The revolution that didn't happen. *New York Review of Books*, vol. 45.

Weiner, J. (1999) *Time, Love, Memory*. Vintage Books, New York.

Welsh, D. K., Logothetis, D. E., Meister, M. & Reppert, S. M. (1995) Individual neurons dissociated from rat suprachiasmatic nucleus express independently phased circadian firing rhythms. *Neuron*, **14**, 697–706.

Went, F. W. (1960) Photo- and thermoperiodic effects in plant growth. *Cold Spring Harbor Symposia on Quantitative Biology: Biological Clocks*. Long Island Biological Association, The Biological Laboratory, Cold Spring Harbor, L.I., New York., pp. 221–30.

Whitehead, D. C., Thomas, H., Jr. & Slapper, D. R. (1992) A rational approach to shift work in emergency medicine. *Ann Emerg Med*, **21**, 1250–58.

Whitmore, D., Foulkes, N. S. & Sassone-Corsi, P. (2000) Light acts directly on organs and cells in culture to set the vertebrate circadian clock. *Nature*, **404**, 87–91.

WHO (1992) World malaria situation 1990. Division of Control of Tropical Diseases. World Health Organization, Geneva. *World Health Stat Q*, **45**, 257–66.

Wilson, E. O. (1998) *Consilience*. Alfred A. Knopf, New York.

Woodrow, H. (1951) Time perception. In Stevens, S. S. (ed.) *Handbook of Experimental Psychology*. John Wiley, New York, pp. 1224–36.

Wright, J. E., Vogel, J. A., Sampson, J. B., Knapik, J. J., Patton, J. F. & Daniels, W. L. (1983) Effects of travel across time zones (jet-lag) on exercise capacity and performance. *Aviat Space Environ Med*, **54**, 132–37.

Wurtman, R. J. & Wurtman, J. J. (1989) Carbohydrates and depression. *Sci Am*, **260**, 68–75.

Yamazaki, S., Numano, R., Abe, M., Hida, A., Takahashi, R., Ueda, M., Block, G. D., Sakaki, Y., Menaker, M. & Tei, H. (2000) Resetting central and peripheral circadian oscillators in transgenic rats. *Science*, **288**, 682–85.

Young, J. Z. (1962) *The Life of the Vertebrates*. Clarendon Press, Oxford.

Young, M. (1998) *The Metronomic Society: Natural Rhythms and Human Timetables*. Thames & Hudson, London.

Zimmerman, N. H. & Menaker, M. (1979) The pineal gland: a pacemaker within the circadian system of the house sparrow. *Proc Natl Acad Sci USA*, **76**, 999–1003.

INDEX